DARWINISM DEFEATED?

Contents

Contributors   5
Foreword by J. I. Packer   7

THE JOHNSON-LAMOUREUX DEBATE

Evangelicals Inheriting the Wind:
The Phillip E. Johnson Phenomenon   9
    Denis O. Lamoureux

Response to Denis O. Lamoureux   49
    Phillip E. Johnson

The Gaps Are Closing:
The Intellectual Evolution of Phillip E. Johnson   57
    Denis O. Lamoureux

Final Response to Denis O. Lamoureux   77
    Phillip E. Johnson

RESPONSES TO THE DEBATE

Intelligent Design: The Celebration of Gifts Withheld?   81
    Howard J. Van Till

Teleological Evolution: The Difference it Doesn't Make   91
    Stephen C. Meyer

Comments on Denis Lamoureux's Essays   103
    Michael J. Behe

Design and Purpose within an Evolving Creation   109
    Keith B. Miller

On Being and Becoming:
Conflation and the Confusion of the 'Science' of Evolution   121
    Michael W. Caldwell

The Intelligent Design Movement,
Evangelical Scientists, and the Future of Biology   137
    Jonathan Wells

The Intelligent Design Movement
Comments on Special Creationism   141
    Michael J. Denton

Of Apples and Star Trek, Guidance and Gaps   155
    Rikki E. Watts

Does Methodological Naturalism lead to
Metaphysical Naturalism?   166
    Loren Wilkinson

Contributors

Michael J. Behe (PhD, University of Pennsylvania) is Associate Professor of Biochemistry, Department of Biological Sciences, Lehigh University, Bethlehem, Pennsylvania, U.S.A.

Michael W. Caldwell (PhD, McGill) is a Research Scientist in the Paleobiology Group at the Canadian Museum of Nature, Ottawa, Canada, and Adjunct Research Professor, Earth Sciences Department, Carleton University, Ottawa, Canada.

Michael J. Denton (PhD, London) is Senior Research Fellow in Human Genetics in the Department of Biochemistry at the University of Otago, Dunedin, New Zealand.

Phillip E. Johnson (J.D., University of Chicago) is Jefferson E. Peyser Professor of Law at the University of California, U.S.A.

Denis O . Lamoureux (DDS, PhD, Alberta; PhD Toronto) lectures on the relationship between science and religion at St. Joseph's College, University of Alberta, Edmonton, Canada.

Stephen C. Meyer (PhD, Cambridge) is the C. Davis Weyerhauser Research Fellow in the Philosophy of Biology at the Discovery Institute in Seattle and an Associate Professor of Philosophy at Whitworth College in Spokane, Washington, U.S.A.

Keith B. Miller (PhD, Rochester) is a Research Assistant Professor in Geology at Kansas State University, Manhattan, Kansas, U.S.A.

J. I. Packer (DPhil, Oxford) is Board of Governors' Professor of Theology at Regent College, Vancouver, B.C., Canada.

Howard J. Van Till (PhD, Michigan State) is Professor Emeritus of Physics and Astronomy at Calvin College, Grand Rapids, Michigan, U.S.A.

Rikki E. Watts (PhD, Cambridge) is Associate Professor, New Testament, at Regent College, Vancouver, B.C., Canada.

Jonathan Wells(PhD, Yale; PhD, Berkeley) is a post-doctoral biologist in the Department of Molecular and Cell Biology, University of California at Berkeley, U.S.A. and a senior research fellow at the Discovery Institute in Seattle, Washington, U.S.A.

Loren Wilkinson (PhD, Syracuse) is Professor of Interdisciplinary Studies and Philosophy at Regent College, Vancouver, B.C., Canada

Foreword

Here is a contribution to a currently expanding discussion in theology and apologetics. Two connected questions are involved. First, is any form of the Bible-based Christian concept of creation compatible with any form of the current biology-based concept of evolutionary development of life? Second, if it is, how precisely do you correlate the concepts? A spectrum of views exists. At one end are scientists who believe in evolution, but not in God. Some of these are agnostic; some think, as did so many in the first half of the century, that the various sciences, put together, have between them managed to disprove all forms of belief in God; all believe that the integrity of science suffers if you try to fit its findings into a theistic, deistic, pantheistic or panentheistic frame. (Polytheism would be mentioned in that list, too, if it was still a live option among religious people.) At the other end of the spectrum are believers in God who do not believe in evolution; these vary among themselves in the way they conceive evolution and understand the biblical witness to God's work of creation. Between the two extremes are many, professional scientists and theologians as well as men and women in the street, who believe in both God and evolution, seeing evolution as one element in God's way of making and ordering his world; and some of these think that natural processes provide evidence of intelligent design, and so reinforce the classic cosmological and teleological arguments for the reality of a rational Creator. There is plenty to talk about here, and the only certainty is that the last word has not been spoken yet.

In the following pages we watch an evolutionary creationist critiquing an anti-Darwinian who evidently believes in intelligent design. The anti-Darwinian strikes back, and reinforcements are brought up on both sides. The seriousness, vigour, rigour, and basic good-will of the disputants are admirable; some of the exchanges are both blunt and sharp, as is often the case in academic discussion, but if the clash of minds stimulates thought, as I think it does here, it is really no bad thing.

A glum comment in a magazine about science-and-religion dialogues declared: "Fundamental differences in definitions of key terms, in background assumptions, and so on result in frustrating exercises in which opposing parties simply talk past each other. Following many of these debates is like trying to grab a web bar of soap." The protagonists in the present in-house discussion are both conservative Christians, labouring to hear each other as well as to address each other, and I think their exchanges, plus the additional material deployed by their respective supporters, will be found genuinely clarifying. Regent College serves the Christian world well in making it all available in print.

J. I. Packer
Regent College

Evangelicals Inheriting the Wind: The Phillip E. Johnson Phenomenon

Denis O. Lamoureux

It is indeed a privilege and a pleasure to review *Defeating Darwinism*[1] and interact with its author, Dr. Phillip E. Johnson, whom I consider the most important antievolutionist in the world today. Even though I believe Professor Johnson's understanding of biological evolution is seriously flawed, he is very correct in identifying the abuse of a certain version of this scientific theory to support a godless worldview. Despite our differences, I know well of his commitment to our Lord and Saviour Jesus Christ, and I would be the first to pass and receive the communion cup to and from him on a Sunday morning. It is my hope and prayer that our exchange will demonstrate the truth of the proverb "As iron sharpens iron, so one man sharpens another"(Prov. 27:17).

Defeating Darwinism is Johnson's third book dealing with the debate over biological origins and its philosophical and theological implications. In this work he compares the modern evolution–creation controversy to the play *Inherit the Wind* which was made into a movie with the same title in 1960 and starred Spencer Tracy, Gene Kelly and Frederic Mach. The story is a fictionalized version of the Scopes Trial, the legendary Tennessee trial in 1925 that convicted a schoolteacher for introducing the theory of evolution to his students. According to Johnson, "*Inherit the Wind* is a masterpiece of propaganda, promoting the stereotype of the public debate about creation and evolution that gives all the virtue and intelligence to the Darwinists."[2] He claims that those who succumb to

---

[1] Phillip E. Johnson, *An Easy-To-Understand Guide for Defeating Darwinism by Opening Minds* (Downers Grove, Ill.: InterVarsity Press, 1997).

[2] Johnson, *Defeating Darwinism*, 25.

this propaganda in the end only inherit the wind. But I believe that Johnson's so-called "inherit-the-wind syndrome" can work both ways.[3] Could it be that his books are a masterpiece of propaganda, promoting the stereotype of the evangelical church's debate about creation and evolution that gives all the virtue and intelligence to Johnson and his antievolutionist colleagues? More specifically, is the current popularity of Johnson's antievolutionism in North American evangelicalism a clear example of this community inheriting the wind? I hope to investigate these questions in this paper.

Before beginning, introductory remarks are required with regard to two topics. First, the terminology employed in the origins debate is confusing and often contradictory, opening the way for a great deal of misunderstanding. For example, terms such as 'evolution', 'Darwinism', and 'creationism' carry a variety of meanings and nuances. They are also emotionally charged and have changed over time in this debate. As a result, I will start with a brief review and define the terms used in this paper. Second, Johnson states that the purpose of *Defeating Darwinism* is to make the ideas he developed in *Darwin on Trial* (1991) and *Reason in the Balance* (1995) accessible to an audience that is "not quite so familiar with university-level subjects."[4] Therefore, before assessing his third book it is necessary to examine and evaluate the foundational principles that Johnson claims to have established in his first two works: (1) the pervasiveness of naturalism and materialism in our society, (2) the evidence for intelligent design in the universe, and (3) the complete failure of the modern theory of biological evolution.

Definition of Terms:
Evolution, Darwinism and Creationism

The terms 'evolution' and 'Darwinism' are powerful and evocative, and their use or misuse has certainly contributed to the confusion in the origins debate. For many evangelicals, these terms are interchangeable and usually associated with a 'molecules to humans' theory driven *only* by blind chance and irrational necessity. In other Christian circles, and among many who believe in the existence of God, evolution is the method

---

[3] "The 'Inherit the Wind' Syndrome" is the title of an audio cassette of a lecture delivered by Johnson in January 1997 at Regent College. It is available from the Regent Bookstore, 5800 University Boulevard, Vancouver, BC, Canada V6T 2E4, 1-800-663-8664; or via the web at <www.regent-bookstore.com>.

[4] *Defeating Darwinism*, 9.

through which God created the universe and the life in it.[5] In other words, there are two radically different uses of the term 'evolution' today. Thus it is necessary to distinguish between: (1) a purposeful and designed process termed *teleological* evolution—the position of some Christians and believers in God, and (2) a purposeless and chance-driven process termed *dysteleological* evolution—the position of atheists, to which all Christians are firmly opposed. Johnson adamantly insists that this second sense of the term is the *exclusive* meaning employed in the scientific community. However, I will argue that he not only overstates the case, but that he is simply wrong.

Second, the term 'Darwinism' is used today not only by Johnson and some Christians in the origins debate to refer to a dysteleological evolution, but also by a number of non-Christian popularizers and philosophers like Richard Dawkins, Daniel Dennett, and Michael Ruse.[6] However, the recent and rapid growth in historical studies on the life of Charles Darwin has made it abundantly clear that his view of evolution was not dysteleological, though he considered that possibility.[7] For example, Darwin acknowledged "the extreme difficulty or rather impossibility of conceiving this immense and wondrous universe, including man with his capacity of looking backwards and far into futurity, as the result of blind chance or necessity."[8] This notion, he admits, was "strong in [his] mind" when he wrote the famed *On the Origin of Species* (1859), but "very gradually, with many fluctuations, [became] weaker."[9] But more

---

[5] See Pope John Paul II's recent pronouncement on evolution suggesting that it is "more than a hypothesis." John Paul II, "Message to Pontifical Academy of Sciences on Evolution," *Origins: CNS Documentary Service* 26 (5 December 1996): 414–416.

[6] See Richard Dawkins, *The Blind Watchmaker* (London: Penguin: New York: W. W. Norton, 1986); Daniel C. Dennett, *Darwin's Dangerous Idea: Evolution and the Meanings of Life* (New York: Simon and Schuster, 1996); Michael Ruse, *Taking Darwin Seriously* (Oxford: Basil Blackwell, 1986). To this list can be added Phillip Johnson, who insists that Darwinism refers to dysteleological evolution.

[7] For examples of these revisionist histories rehabilitating Darwin's theism/teleology, see Adrian Desmond and James Moore, *Darwin: The Life of a Tormented Evolutionist* (New York: Warner, 1991); Neal C. Gillespie, *Charles Darwin and the Problem of Creation* (Chicago and London: University of Chicago Press, 1979); Denis O. Lamoureux, *Between "The Origin of Species" and "The Fundamentals": Toward a Historiographical Model of the Evangelical Response to Darwinism in the First Fifty Years*. PhD dissertation. University of St. Michael's College and Wycliffe College at the University of Toronto, 1991.

[8] Charles Darwin, *The Autobiography of Charles Darwin, 1809–1882*. Edited by Nora Barlow (London: Collins, 1958), 92.

[9] Ibid., 93. In the *Origin of Species* Darwin openly and unapologetically refers to a "Creator" on seven occasions. See Charles R. Darwin, *On the Origin of Species*. Facsimile of the first (1859) edition, introduced by Ernst Mayr (Cambridge, Mass.: Harvard University

importantly, modern professional biologists employ the term 'evolution' with regard to biological origins and rarely use *Darwinism*. In a computer search of titles, abstracts, and subject heads used in the professional literature in the biological sciences between 1992 and mid-1997, the word 'evolution' was employed 29,803 times, while a meagre twelve entries appeared for 'Darwinism'.[10] In other words, those *directly involved* with the biological theory of evolution use the term evolution instead of Darwinism at a rate of nearly 2500 times to 1.

As a result, for the sake of clarity, historical accuracy, and linguistic convention in biology, I do not use the term 'Darwinism' in the modern origins debate. When the word 'evolution' is employed without any qualification, it refers simply to the notion of common descent (i.e., the molecules to humans theory) and does not indicate the overriding nature of the process (i.e., whether purposeful/designed or purposeless/chance-based). The qualifiers teleological or dysteleological are thus used in contexts where it is necessary to specify the character or type of evolutionary belief.

In 1989 Roger E. Timm accurately observed that, "In current usage 'creationist' often is meant to refer to those who believe in the literal truth of the Genesis creation accounts."[11] Since that time there has been a quiet revolution among many North American evangelicals as they have come to accept with little resistance the standard cosmological and geological dating of the universe and earth in the billions of years.[12] In order to dis-

---

Press, 1964), 186, 188, 189, 413 (twice), 435, 488. In the 2nd edition of the *Origin*, published in 1860, Darwin even adds "by a Creator" to the famous last sentence of his book: "There is grandeur in this view of life [i.e., evolution], with its several powers, having been originally breathed *by a Creator* into a few forms or into one; and that, whilst this planet has gone on cycling according to the fixed law of gravity, from so simple a beginning endless forms most beautiful and most wonderful have been, and are being, evolved."

[10] The computer search was done by the Cambridge Scientific Abstracts Internet Database Service at the University of Alberta, Edmonton, Canada. The investigation covered the titles, abstracts, and subject headings of published professional literature in the biological sciences from 1992 to June 30, 1997. A search of the root term 'Darwin_' resulted in over 600 entries. Of these, 490 were simply the proper name 'Darwin.' 'Darwinian' appeared 106 times, always in the context of the Darwinian mechanism of selection.

[11] Roger E. Timm, "Scientific Creationism and Biblical Theology," in *Cosmos as Creation: Theology and Science in Cosmology*, ed. Ted Peters (Nashville: Abington Press, 1989), 260.

[12] I cannot help but speculate that the worldwide popularity of Stephen Hawking's *A Brief History of Time: From the Big Bang to Black Holes* (New York: Bantam, 1988) has been a significant factor in the quiet acceptance of an old universe amongst well-read evangelicals. In addition, the works of evangelicals Davis Young (geologist) and Hugh Ross (astronomer) have also influenced these evangelicals in coming to terms with the age of the universe. See Davis A. Young, *Christianity and the Age of the Earth* (Grand Rapids: Zonder-

tinguish the previous view of creation from this newly emerging position, it has become common to hear the term 'young earth creationism' used to describe the views of those holding a very strict, literal interpretation of Genesis. This position maintains that the earth is only about 10,000 years old, that the creation of life in its fully developed form happened in six 24-hour days, and the worldwide flood of Noah accounts for most geological stratification.

The term 'old earth creationism' is used for the new position, but many also refer to it as 'progressive creationism'.[13] Progressive creationists claim that God created in a progressive manner over the course of geological time through direct intervention because the universe is incomplete and life does not have the potential to evolve. The exact nature of this intervention is a matter of debate. Some suggest that completely new species were created in different geological epochs, others contend that genetic material was added to and/or manipulated in pre-existing organisms to produce new species, and many uphold a combination of all these methods. Most progressive creationists accept a certain plasticity in biological life that allows for some evolutionary change, even into higher forms of life. However, they contend there are limits to this biological plasticity, and as a result gaps or discontinuities exist in the history of life requiring the direct intervention of God. Most progressive creationists are also biblical concordists. That is, they suggest that there is a general correlation between the early chapters of the Bible and modern science. For example, some maintain that the days of Genesis relate to geological ages and that Noah's flood corresponds to a historical diluvial event limited to the Mesopotamian valley. Phillip Johnson's views on biological origins clearly fall within the progressive creationism camp. Though he conscientiously avoids making any direct statements about how the Bible relates to his view of origins, a few passages in his works clearly suggest that he upholds a form of concordism.

The last two terms requiring definition are 'evolutionary creationism' and 'theistic evolution'. Both views claim that God created through an

---

van, 1982); Hugh Ross, *Genesis One: A Scientific Perspective* (Pasadena, Cal.: Reasons to Believe, 1983); *The Fingerprint of God* (Orange, Cal.: Promise, 1991 [1989]); *Creation and Time: A Biblical and Scientific Perspective on the Creation-Date Controversy* (Colorado Springs: NavPress, 1994).

[13] Michael Ruse recognises this division within evangelicalism and differentiates between the old antievolutionism (young earth creationism) and the new antievolutionism (progressive creationism). Michael Ruse, "The New Antievolutionism," transcribed by Paul A. Nelson. *Meeting of the American Association for the Advancement of Science*, 1993.

evolutionary process. The difference between these positions is theological. Evolutionary creationists are distinctly conservative Christians,[14] while the theology of theistic evolutionists ranges from liberal Christianity to philosophical theism (i.e., a belief in God based on philosophical reasoning). Process theology often plays an important part in the theology of theistic evolutionists. The position I hold is evolutionary creationism.[15]

To conclude, it is clear that the word 'creationism' carries a wide variety of meanings, and clarification of its meaning is necessary. However, if it is to be used without qualification, then Timm's proposal to employ it with a wider sense must be considered:

> The intent of my argument here is to reclaim the title 'creationist' for those who willingly confess that God is creator but who do not necessarily interpret the Genesis creation accounts literally [or attempt a concordism with modern science]. My claim is that being a creationist means affirming the theological implications of the biblical creation accounts. A creationist by this standard is someone who believes in a creator who is one, good, and caring, who created the world good, and who has given humans responsibility for caring for God's creation and for living in harmony with it.[16]

---

[14] Regarding the term 'evolutionary creationism', see Timm, "Scientific Creationism and Biblical Theology," 261–262. Important works by evangelical evolutionary creationists include Howard J. Van Till, *The Fourth Day: What the Bible and the Heavens are Telling Us about the Creation* (Grand Rapids: Eerdmans, 1986); Howard J. Van Till, ed. *Portraits of Creation: Biblical and Scientific Perspectives on the World's Formation* (Grand Rapids: Eerdmans, 1990).

[15] It is important to distinguish evolutionary creationism from deism because Johnson and many of his supporters suggest these two positions are essentially the same. Deism is classically understood as the view that God created the universe and then turned his back on it never to be active in it (i.e., no miraculous intervention) nor to reveal himself in it (i.e., no revelation through prophets or prayer). This is like winding a clock and leaving it alone to wind down. In contrast, evolutionary creationism holds that God created through physical laws that he ordained and continues to sustain. It is because of a regularity of phenomena in nature that science can describe these patterns and call them laws. In addition, evolutionary creationism acknowledges God's direct activity in his creation through miraculous signs and wonders and the revelation of his will through prophets and Jesus Christ. Quipping with my colleagues, I often describe myself as a "signs and wonders evolutionary biologist" since that is indeed my experience in science and my faith walk.

[16] Timm, "Scientific Creationism and Biblical Theology," 260. Bracketed words are my inclusion, with which I believe Timm would agree.

The Foundational Principles of Johnson's Position

Principle #1: The Pervasiveness of Naturalism and Materialism

Johnson states that the "unofficial religion" of the modern world as reflected in science, law, and education is naturalism or materialism. He uses these terms interchangeably and defines them as the worldview that sees reality consisting of only "the fundamental particles that make up both matter and energy."[17] In such a world there is no place for a God or gods, or any element of transcendence or a spiritual dimension. Johnson cites examples where a dysteleological interpretation of evolution is used to support this unofficial religion, and these evoke a powerful response in not only Christians, but most people. Indeed, Johnson's review of the 1995 American National Association of Biology Teachers' statement on the character of evolution makes most of us recoil. Specifically, policy-makers in this association conceive evolution dysteleologically as "an unsupervised, impersonal, unpredictable, and natural process."[18]

Johnson's exposition of naturalism/materialism is worthy of serious consideration. In modern society, science is presented as an intellectual and cultural value; Johnson correctly points to examples where it is effectively elevated to the status of a religion. But in fairness to those before Johnson, this is not an original observation because in this century both Roman Catholics and many evangelicals (e.g., American Scientific Affiliation) have long recognized that science can be transformed into an atheistic worldview often termed scientism or scientific materialism. Even secular philosophers like Michael Ruse have argued against this misuse of science.[19]

However, Johnson's claim that naturalism/materialism has pervaded modern society must be challenged. For example, is the scientific community as thoroughly naturalistic and materialistic as he declares? Regarding biological origins, is Johnson correct in asserting that contemporary scientists and science educators "are *absolutely insistent* that evolution is an unguided and mindless process, and that our existence is

---

[17] *Defeating Darwinism*, 15. See also Phillip E. Johnson, *Reason in the Balance: The Case Against Naturalism in Science, Law and Education* (Downers Grove, Ill.: InterVarsity Press, 1995), 38.

[18] *Defeating Darwinism*, 15. Quoted by Johnson from the National Association of Biology Teachers, "Statement on Teaching Biology," *American Biology Teacher* 58 (1996): 61–62.

[19] Recently Ruse has even asserted (to the surprise of many) that science is akin to religion in that it requires making certain philosophic assumptions, which at some level cannot be proven scientifically. Ruse, "The New Antievolutionism."

therefore a fluke rather than a planned outcome"?[20] In direct opposition to Johnson, Edward Larson and Larry Witham reported in the prestigious scientific journal *Nature* (3 April 1997) that four out of ten leading American scientists believe in a *personal* God.[21] Their paper, entitled "Scientists Are Still Keeping the Faith," found that this ratio is identical to that found in a duplicate survey performed by James Leuba in 1916.[22] Impetus for the recent study was to test the prediction made by Leuba that with the spread of education disbelief would increase. However, comparing the two surveys shows that little has changed in eighty years regarding belief in a personal God. As a result, the gloomy picture with near conspiratorial tones that Johnson paints with regard to the pervasiveness of naturalism/materialism is not as dark, sinister, and widespread as he claims because this is certainly not the case in the American scientific community today.

To be sure, during this century the biological theory of evolution has come to be the only paradigm for the origin of life in the scientific world. However, in the light of the Larson and Witham study, it does not follow that to be an evolutionist one is *necessarily* a naturalist or materialist denying the existence of God, as Johnson constantly insists in his books. Rather, it is reasonable to suggest that at least four out of ten scientists in the U.S. believe God created through a *teleological* evolutionary process, and the belief in this view of evolution is probably much higher.[23] The so-called 'believers' in this study were limited to those who held a narrow view of the Divine Being as a *personal* God. Unfortunately, scientists who

---

[20] *Defeating Darwinism*, 15. My italics.

[21] This study was based on a random selection of 1000 leading American scientists from the current edition of *American Men and Women in Science*. In considering the belief in a personal God, the scientists in both studies were asked to evaluate the following statement: "I believe in a God in intellectual and affective communication with humankind, i.e., a God to whom one may pray in expectation of receiving an answer. By 'answer' I mean more than the subjective, psychological effect of prayer." In 1997, 40 percent of the scientists agreed that this statement was true; 45 percent of them asserted that they "do not believe in a God as defined [in the statement];" and 15 percent of the responding scientists had "no definite belief regarding this question." Edward J. Larson and Larry Witham, "Scientists Are Still Keeping the Faith" *Nature* 386 (3 April 1997): 435–456.

[22] James H. Leuba, *The Belief in God and Immortality: A Psychological, Anthropological and Statistical Study* (Boston: Sherman, French and Co., 1916). Larson and Witham conclude, "Though a noted psychologist, Leuba misjudged either the human mind or the ability of science to satisfy all human needs." Larson and Witham, "Keeping the Faith," 435.

[23] In making this suggesting, I am assuming that nearly all the scientists in the study are evolutionists (i.e., they accept common descent). This is an assumption I believe Johnson would accept.

had other conceptualizations of God, and who thus also accepted teleological evolution, were grouped with respondents who completely de-
nied the existence of God (this so-called 'disbelievers' group accounted
for 45 percent of scientists). In addition, 15 percent of those polled were
agnostic on the existence of a personal God. As a result, a rough calculation would suggest that the term evolution carries a dysteleological sense
for less than half of the scientists surveyed, which is far from Johnson's
view that contemporary scientists and science educators "are absolutely
insistent that evolution is an unguided and mindless process."

A second landmark paper entitled "Science and God: A Warming
Trend?" has been published in *Science*, the journal of the prestigious
American Association for the Advancement of Science. In it Gregg
Easterbrook notes:

> Both the National Academy of Sciences and the American Association
> for the Advancement of Science . . . have launched projects to promote a
> dialogue between science and religion. New institutions aimed at
> bridging the gap have been formed, including the Chicago Center for
> Religion and Science and the Center for Theology and Natural Sciences
> in Berkeley, California. Universities such as Cambridge and Princeton
> also have established professorships or lectureships on the reconcilia-
> tion of the two camps.[24]

This paper also offers numerous examples of first-rate scientists who are
devoutly religious. Easterbrook closes by suggesting that "Rather than
being driven even farther apart, tomorrow's scientist and theologian may
seek each other's solace."[25]
In sum, Johnson is to be congratulated for pointing out blatant examples
where materialism and naturalism are injudiciously expressed and imposed in certain sectors of our society. However, he overstates his case
with regard to the pervasiveness of this philosophical view, and he is
simply wrong in suggesting that materialism/naturalism is *necessarily*
associated with the biological theory of evolution or that this dysteleological worldview is universally upheld by the modern scientific
community.

Principle #2: The Evidence for Intelligent Design in the Universe

Johnson maintains that the intricate and complex design manifested
in the universe is evidence of an intelligent designer. In the last ten or so

---

[24] Gregg Easterbrook, "Science and God: A Warming Trend?" *Science* 277 (15 August
1997): 890.
[25] Ibid., 893.

17

years, a loosely defined group known as the intelligent design theorists has appeared in American evangelical circles. Their most important publications include *Of Pandas and People* (1989), *The Creation Hypothesis* (1994), *Darwin's Black Box* (1996), and of course, Johnson's *Darwin on Trial* (1991).[26] This movement had its first international conference in late 1996 with Johnson as its featured speaker. The conference attracted over two hundred scientists and scholars who "reject naturalism as an adequate framework for doing science, and who advocate a common vision of creation whose unifying idea is intelligent design."[27]

Evidence for the powerful impact that design in nature has on the human spirit is seen throughout the history of ideas. The notion that this design points to a designer is not a distinctly Christian idea but transcends all cultures and times — from Hebrew psalmists to Greek philosophers to twentieth-century physicists. The so-called "argument from design" is one of the most powerful and widely used defences for the existence of God in church history. The Word of God emphatically confirms this view:

> The heavens declare the glory of God;
> the skies proclaim the work of his hands.
> Day after day they pour forth speech;
> night after night they display knowledge.
> There is no speech or language
> where their voice is not heard.
> Their voice goes out into all the earth,

---

[26] Percival Davis and Dean Kenyon, *Of Pandas and People: The Central Question of Biological Origins* (Dallas: Haugthon, 1989); J. P. Moreland, ed., *The Creation Hypothesis: Scientific Evidence for an Intelligent Designer* (Downers Grove, Ill.: InterVarsity Press, 1994); Michael J. Behe, *Darwin's Black Box: The Biochemical Challenge to Evolution* (New York: Free Press, 1996).

[27] This is the conference purpose as cited in the advertisement for "Mere Creation — Reclaiming the Book of Nature: Conference on Design and Origins," Biola University, November 14–17, 1996. A survey of the participants regarding their views on origins was taken, but has yet to be released. I attended this conference and would not be surprised if nearly one-quarter of the participants were teleological evolutionists. The majority were progressive creationists and only a few young earth creationists were present. Also in attendance was Michael Denton who has been an important influence in the origins debate during the last fifteen years. To his chagrin, North American evangelicals and fundamentalists have misrepresented his views. He does concede, however, that his imprecise use of term 'evolution' in his writings has contributed to the problem. Denton admitted to me that he has long accepted a teleological view of evolution and adamantly dismisses the dysteleological version. See Michael Denton, *Evolution: A Theory in Crisis* (Bethesda: Alder and Alder, 1986). Denton plans another edition of this book and hopes to correct this oversight.

their words to the ends of the world.
(Ps. 19: 1–4)

Furthermore, St. Paul asserts that humanity can even know "God's invisible qualities—his eternal power and divine nature" because they are "plain" to us, "being understood from what has been made" (Rom. 1:18–20). The apostle concludes that this revelation through creation is so clear that it judges humanity because we are accountable for its message.

Testifying to the power and reality of design in the universe is the fact that as time and scientific investigation have advanced, greater manifestations of nature's intricacy and complexity have emerged. This only makes a stronger argument for intelligent design. Johnson and the design theorists, however, introduce a unique twist to the notion of design. For them, design carries an aspect of irreducible complexity. That is, they assert that certain biological structures are fashioned in such a way that it was not possible for them to develop through a natural process like evolution (whether teleological or dysteleological). To account for the existence of these irreducibly complex structures, intervention from outside the normal operation of the universe is claimed to have occurred during the history of life. As a result, the design theorists are progressive creationists. Such a position, however, leaves itself open to criticism for being another version of the God-of-the-gaps. That is, once natural processes are discovered to account for the creation of a once acclaimed irreducibly complex structure, God's purported intervention is lost to the advancing light of scientific research. A serious consequence of filling these gaps (once believed to be the sites of God's active hand) is that God appears to be forced further and further into the dark recesses of our ignorance; and yes, the dangerous notion arises that maybe human ignorance is in effect the 'creator,' a resident only of our minds.

Johnson then correctly affirms the time-honoured notion that the universe reflects intelligent design. However, he should be cautious not to suggest that the establishment of this design *necessarily* requires direct divine intervention because natural processes ordained and sustained by the Creator could account for the universe and life in its God-glorifying splendour.

Principle #3: The Complete Failure of the Modern Theory of Biological Evolution

Johnson believes that the modern theory of biological evolution (or Darwinism, the term that he prefers to employ) is hopelessly flawed with regard to both the evidence supporting the theory and the logic employed to argue for it. This third principle has gained Johnson international recognition, and as a result he has rapidly emerged as an intellectual leader in North American evangelicalism.

In *Defeating Darwinism*, Johnson claims that he "had taken on the scientific evidence" for biological evolution in his earlier work, *Darwin on Trial*.[28] Being a professor of law who steps outside his field of expertise to challenge the theory that is the unifying paradigm of the entire biological community, he justifies entering this discussion by appealing to his academic qualifications:

> I am not a scientist but an academic lawyer by profession, with a specialty in analyzing the logic of arguments and identifying the assumptions that lie behind those arguments. This background is more appropriate than one might think, because what people believe about evolution and Darwinism depends very heavily on the kind of logic they employ and the kind of assumptions they make.[29]

Johnson's conclusion after applying his "specialty in analyzing the logic of arguments" is that the theory of evolution falls hopelessly short. Of course, a corollary that can be gleaned from his judgment is that professional biologists worldwide are simply poor thinkers. This is an interesting and even frightening thought since most health professionals first begin their university training in departments of biology—more later regarding medical science and Johnson.

However, everyone will agree that before Johnson can use his analytical tools, he must first demonstrate a solid grasp of the fundamentals of biology and the evidence for evolution. More specifically, he has to have a firm knowledge of the fossil record for it is upon this foundational evidence that the theory of evolution was first constructed and claims to remain a viable hypothesis. Therefore, let us examine Johnson's evaluation of some of the best evidence for evolution, the fossil record of vertebrates.

In considering this evidence, he admits in *Darwin on Trial* that "the primary source for the information about the vertebrate fossil record in this chapter [i.e., chapter 6: "The vertebrate sequence"] is Barbara J. Stahl's comprehensive text *Vertebrate History: Problems in Evolution* (Dover, 1985), especially chapters five and nine."[30] But a quick review of the

---

[28] *Defeating Darwinism*, 9.
[29] Phillip E. Johnson, *Darwin on Trial* (Downers Grove, Ill.: InterVarsity Press, 2nd edition, 1993 [1991]): 13.
[30] *Darwin on Trial*, 189–190.

scholarship that Stahl employs in these two chapters reveals that 85 percent of her references are prior to 1970, and only 4 percent of the citations are between 1980 and 1982, these being the *latest* sources she uses.[31] More importantly, Stahl's book has had little impact in the scientific community. From the time it was first published in 1974 until 1989, it was cited a mere twenty-seven times in the scientific literature.[32] In the period between 1990 and 1996, the time when Johnson was writing and publishing his two editions of *Darwin on Trial* (1991 and 1993), only two citations appeared. In other words, the primary source that Johnson used to build one of his most important arguments against the theory of evolution is *one* outdated introductory textbook that has received little attention in the scientific community.[33]

To further appreciate Johnson's grasp of the fundamentals of biology and evolutionary theory, let us consider his understanding of the evolution of whales. He asks:

> By what Darwinian process did useful hind limbs wither away to vestigial proportions, and at what stage in the transformation from rodent to sea monster did this occur? Did rodent forelimbs transform themselves by gradual adaptive stages into whale flippers? We hear nothing of the difficulties because to Darwinists unsolvable problems are not important.[34]

Considering Johnson's last comment, he forgets that he has admitted that his knowledge of the vertebrate fossil record is based on Stahl's book, a work that is subtitled "*Problems* in Evolution" [my italics]. Her text is clear evidence that evolutionists openly admit that there are difficulties with the theory, and that they are interested in grappling with these problems openly in the literature. There is no scientific discipline or theory that does not have problematic areas, and evolutionary biology is no exception.

---

[31] It would appear that the second edition of Stahl's book (1985) is more of a reprint than a new edition. In the chapters in question (chapters 5 and 9) the bibliography has increased by only four and two references, respectively. There seems to be a decade of scholarship since the first edition (1974) with which she has not interacted.

[32] *Science Citation Index* (Philadelphia: Institute for Scientific Information). In the period between 1975 and 1984 Stahl's book received nineteen citations. In contrast, the most respected paleontology textbook at that time was A. S. Romer's *Vertebrate Paleontology* (1966) which was cited in the scientific literature 587 times.

[33] I hope that Johnson does not suggest that the poor reception of Stahl's book is because it was interpreted to be an antievolutionary work. Stahl is thoroughly committed to evolution and has the integrity to underline problems with the theory.

[34] *Darwin on Trial*, 87.

More importantly, Johnson's biology is displayed in the preceding passage. First, no modern evolutionary biologist believes that whales descended from rodents (for that matter, I cannot think of anyone who ever did). Rather, scholarly consensus today suggests that whales evolved about fifty-five million years ago from mesonychid ungulates (ungulates are mammals with hoofs) that were a specialized group of carnivores (i.e., animals with teeth for tearing and slashing flesh). If Johnson had consulted the two most important and basic textbooks of vertebrate paleontology in our generation—Alfred Sherwood Romer's *Vertebrate Paleontology* (1966) and Robert L. Carroll's *Vertebrate Paleontology and Evolution* (1988)—he would not have made such an serious error.[35] For that matter, if Johnson had referred to a work by his antievolutionary colleagues, *Of Pandas and People* (1989), this mistake would never have occurred.[36] Similarly, he should have read Stahl's book with more care because she makes no reference to the evolution of whales from rodents. Instead, she emphasizes the theory of the mesonychid origin of whales.[37] I believe Johnson's rodent-to-whale passage is very significant and clearly reveals the law professor's limited and erroneous understanding of the data used to support evolutionary theory.

Second, Johnson finds it incredible that legs could give way to flippers during the evolution of whales. But a basic knowledge of modern developmental biology (also called *embryology*) quickly solves this problem. The experimental manipulation of a region that controls the development of the limb (called the zone of polarizing activity) or modification of the expression of the gene associated with this controlling region in the developing limb (the segment polarity gene called 'Sonic hedgehog') can give remarkable variation in the anatomy of the bones in the

[35] Alfred Sherwood Romer *Vertebrate Paleontology*, 3rd ed. (Chicago: University of Chicago Press, 1966 [1933], 297–301. Robert L. Carroll, *Vertebrate Paleontology and Evolution* (New York: Freeman, 1988), 520–521.

[36] Davis and Kenyon, 101. I find it remarkable that Johnson's colleagues and supporters have not pointed out this error to him after its appearance in the first edition of *Darwin on Trial* (1991, 85) since the passage remains uncorrected in the second edition (1993, 87).

[37] See Stahl, *Vertebrate History*, 487. I cannot help but speculate on the origin of Johnson's rodent-to-whale hypothesis. In one brief sentence Stahl refers to a time early in the theorizing of whale evolution when "workers tended to look toward the insectivore-creodont assemblage as likely ancestral stock" because ancient whales were carnivorous. *Ibid.* However, insectivores are small insect-eating mammals like shrews, hedgehogs, and moles. Johnson apparently does not recognize that they form an entirely distinct order from the order of rodents, and he probably thought they were identical. Creodonts are ancient terrestrial carnivores that are the ancestors of the modern order of Carnivores.

limb. Similarly, minor changes in the timing of a series of genes (special controller genes called 'Hox genes' — the series D9 to 13 and A10 to 13) that are expressed during limb development could also conceivably result in dramatic morphological change.[38]

Finally, Johnson gives the impression in *Darwin on Trial* that there is only one dubious transitory form between land mammals and whales — *Basilosaurus*. In evaluating the evidence of the fossil record he claims:

> That 130 years of very determined efforts to confirm Darwinism have done no better than to find a few ambiguous supporting examples is significant negative evidence. . . . The fossils provide much more discouragement than support for Darwinism when they are examined objectively, but objective examination has rarely been the object of Darwinist paleontology. The Darwinist approach has consistently been to find some supporting fossil evidence, claim it as proof for 'evolution,' and then ignore all the difficulties. The practice is illustrated by the use that has been made of a newly discovered fossil of a whale-like creature called *Basilosaurus*. . . . Accounts of the fossil in the scientific journals and in the newspapers present the find as proof that whales once walked on legs and therefore descended from land animals. [39]

It must first be explained that the debate concerning a recent fossil find of this ancient whale is that small legs, according to its discoverers, "were found in direct association with articulated skeletons."[40] Previous specimens featured only reduced hip bones.[41] Johnson's antievolutionary inclination is seen in this passage because he is suspicious of the evidence collected and the conclusion that these reduced legs actually belong to the whale. However, he fails to deal with the fact that previous specimens of *Basilosaurus* (and other ancient whales) had small hip bones which clearly suggest that these whales or their nearest ancestors had

---

[38] It appears that the timing of the expression of the Hox genes (termed 'the Hox combinatorial code') within a developmental field like the limb is a significant factor in determining morphology. For an introduction to the exciting and rapidly developing field of evolutionary developmental biology where these theories are being formulated see Rudolf A. Raff, *The Shape of Life: Genes, Development, and the Evolution of Animal Form* (Chicago: University of Chicago Press, 1996). Primary developmental literature with evolutionary implications for the limb includes: *Journal of Embryology and Experimental Morphology* 87 (1985): 163–174; *Developmental Biology* 109 (1985): 82–95; *Nature* 342 (1989): 767–772; *Nature* 358 (1992): 236–239; *Development* 116 (1992): 289–296; *Nature* 361 (1993): 692–693.

[39] *Darwin on Trial*, 86. My italics.

[40] P. D. Gingerich, B. H. Smith, and E. L. Simons, "Hind Limbs of Eocene Basilosaurus: Evidence of Feet in Whales," *Science* 249 (15 July 1990): 154–157.

[41] R. Kellogg, "A Review of the Archaeoceti," *Carnegie Institute Washington Publications* 482 (1936): 1–366.

small vestigial legs. Predictability is a fruit of a good scientific theory. In this case this is exactly what the theory of evolution predicted (i.e., that legs would be found) and new fossil evidence confirmed it at a later date. Interestingly, a similar fossil finding of a snake with hind legs has recently been reported by Michael Caldwell, one of my colleagues at the University of Alberta.[42] *Pachyrhachis* is a primitive aquatic snake about one metre long with a well-developed pelvis and a pair of small legs a few centimetres in length. It appears that snakes descended from aquatic reptiles that over time in the oceans lost their legs in a similar fashion to the ancestors of whales. Thus there exists evidence in two separate classes of vertebrates for the reduction and even loss of limbs over geologic time.

Johnson does not seem to be aware of the many fossils that make up the whale evolutionary series.[43] There are remarkable jaw and dental similarities between the land carnivore *Hapalodectes* (an early Eocene[44] mesonychid) and *Pakicetus* (late Eocene), one of the oldest known whales. Interestingly, the remains of this ancient whale are also found together with terrestial animals, suggesting that it spent some of its life on land. Moreover, limb reduction is clearly seen during the Eocene from *Ambulocetus natans*, a whale with stubby front legs and well-developed hind legs, through *Rodhocetus* with its powerful tail and hind legs a third the size of *A. natans*, to the small-legged whales *Basilosaurus*, *Protocetes*, and *Zygorhiza* that could no longer walk on land. Throughout this fossil series there is a gradual change in the dentition from carnivorous, multi-rooted, complex teeth to single rooted, cone-shaped teeth, again clearly supportive of evolution.[45]

---

[42] Michael W. Caldwell and Michael S. Y. Lee, "A Snake with Legs from the Marine Cretaceous of the Middle East," *Nature* 386 (17 April 1997): 705–709.

[43] I cannot help but speculate the reason Johnson believes there are so few intermediate fossils in the whale series is because his knowledge is (as he admits) based on Stahl's book. But it is obvious that Stahl's intention was not to give an exhaustive account, especially since she devotes a mere four paragraphs to this topic.

[44] The Eocene period is between 55 and 38 million years ago.

[45] I must firmly emphasize that I am offering only the briefest of sketches of whale evolution and am in no way whatsoever attempting to give a full account of this complex and extensive subject. My point is simply that Johnson does not seem to be aware of the most basic *textbook* knowledge of whale paleontology. For an introduction to this topic, see Carroll, *Vertebrate Paleontology*, 520–527; Annalisa Berta, "What Is a Whale?" *Science* 263 (14 January 1994): 180–181. However, in fairness to Johnson, in *Defeating Darwinism* (p. 60) he does make a passing reference to *A. Natans*, a transitory form that has gained public attention. I cannot help but speculate that it is because this fossil discovery gained media attention that Johnson is aware of it, and not because Johnson is familiar with the primary

To conclude, before Johnson can apply his "specialty in analyzing the logic of arguments and identifying the assumptions that lie behind those arguments" he must first demonstrate at the very least a reasonable grasp of the evidence for evolution and a knowledge of the fundamentals of biology. This brief review of his understanding of whale evolution makes it clear that Johnson is simply not familiar with the topic; and thus the application of his analytical skills and their final results must be deemed suspicious at best, if not outright unacceptable. Johnson bases his views of the vertebrate fossil record on a few chapters in a dated textbook that gained little recognition in the scientific community. He fails to acknowledge the rich whale fossil record (of which I have offered only the briefest of sketches), and his lack of awareness of the basic mechanisms of biological development disables him from speculating on how modification of these processes could easily result in the transition from limbs to flippers during whale evolution. And let the reader be advised, this example is but one small 'biopsy' of Johnson's biology; an entire book could be written responding to its many oversights and misrepresentations. I believe Professor Johnson's understanding of biological evolution is seriously flawed.

Johnson's Foundational Principles and the Problem of Conflation

A significant factor in the acceptance of Johnson's *Darwin on Trial* (1991) and *Reason in the Balance* (1995) in the North American evangelical community relates directly to the way he presents his arguments.[46] His three foundational principles are so tightly interwoven throughout his writings that it becomes nearly impossible for the reader to distinguish them, and this opens the way for the problem of the conflation of ideas. When this happens a poorly rationalized idea can be "justified" simply by being placed alongside a powerful truth. A classic example of this confusion was seen in my Pentecostal church a number of years ago in a presentation by a self-acclaimed end-times specialist. During a series of public lectures on his interpretation of the Book of Revelation, this man also included a magnificent slide presentation of many physical phenomena in

---

scientific literature. For information regarding this ancient whale, see J. G. M. Thewissen, S. T. Hussain, and M. Arif, "Fossil Evidence for the Origin of Aquatic Locomotion in Archaeocete Whales," *Science* 263 (14 January 1994): 210–212.

[46] I would like to emphasize that I do not believe Johnson is consciously aware of this rhetorical move in the sense that he is manipulating his readers with this line of argument.

the universe that clearly underlined that design in nature points strongly toward a Designer. As noted earlier, the argument from design has a great impact on the human spirit and psyche. Consequently, the audience was powerfully moved by images that certainly "declare the glory of God," and the presentation immediately afterward of the speaker's view on the end times allowed these ideas to be carried along in the wake of the design argument's impact and 'authorized' unsuspectingly in the minds of the audience.

This phenomenon of the conflation of ideas is operating in Johnson's writings with regard to his three foundational principles. As I affirmed earlier, with qualification, Johnson's first two principles are powerful and clearly welcomed by all Christians: (1) an attack against naturalism and materialism, (2) support for intelligent design in the universe. However, I caution readers not to conflate these two powerful ideas with Johnson's third foundational principle—the complete failure of the modern theory of biological evolution. In addition, it is important to understand that after Johnson incisively and judiciously exposes the misuse of a *version* of evolution (i.e., dysteleological evolution), he gives the impression that Christians are left with only one option—the acceptance of his antievolutionary biology. Similarly, it must be added that Johnson's use of the powerful design argument, in a fashion similar to the aforementioned end-times speaker, does not force us to accept a view of the origin of life marked by God's direct interventions through geological history. It is logically possible that all the design evident in the universe, which so powerfully testifies to the work of a Creator, could have come about through a God-ordained and sustained evolutionary process (i.e., teleological evolution). In other words, the Creator could well have employed his physical laws and processes to create all the glorious life that we see on this planet today in the same way that his physical laws and processes crafted us in the wombs of our mothers.

A Review of Johnson's *Defeating Darwinism*

Johnson's *Defeating Darwinism* is built on his three foundational principles, but in contrast to his earlier two books he is distinctly more theological and his evangelicalism is openly revealed. As previously noted he has targeted those not familiar with university-level topics in this short book, and I believe he will influence many in that audience. Let us now turn to Johnson's latest book in order to determine if he defeats Darwinism by opening minds as he claims.

Three Common Mistakes: Emilio's or Phil's?

Johnson opens his book with a true story about a devout Christian university student (whom he renames "Emilio") in Europe who claims to have come to terms with the theory of evolution. However, Johnson questions the logic of Emilio's argument and asserts that the student makes three common mistakes.

"Mistake number one," according to Johnson, is that "Emilio is kidding himself about what 'evolution' means. It doesn't mean God-guided, gradual creation. It means unguided, purposeless change. The Darwinian theory doesn't just say that God created slowly. It says that naturalistic evolution is the creator, and God had nothing to do with it."[47] But as noted earlier, the Larson and Witham study shows that 4 out of 10 leading American scientists believe in a personal God. It is fair to assume that the vast majority of these believing scientists are evolutionists because this is the working model accepted across the scientific community. As a result, for at least 40 percent of U.S. scientists the term evolution (which is clearly the word they use, and not Darwinism[48]) does not carry the dysteleological sense that Johnson attempts to impose on it. Here then is *mistake number one* for Johnson: he is simply wrong in asserting that contemporary scientists and modern science educators "are absolutely insistent that evolution is an unguided and mindless process, and that our existence is therefore a fluke rather than a planned outcome."[49]

"Mistake number two," suggests Johnson, is that "Emilio is willing to exchange the Creator of the Bible for the lifeless First Cause of deism. It's like trading real gold for counterfeit money."[50] I am sure that wherever in the world Emilio may be today, he must be frustrated by Johnson's remark. Emilio could argue that, though God ordained and sustained physical laws from the Big Bang to the emergence of humanity bearing his precious image, it is *not logically necessary* to 'retire' God after the initial creation or to confine his activity to that point in the beginning of time. Nor does a teleological evolution dismiss "whether God cares about us" or "whether we need not care about God's purposes."[51] To be sure, Johnson clearly has reason for concern (as I have) about those who limit God's activity only to the initiation of an evolutionary process, and

---

[47] *Defeating Darwinism,* 16.
[48] See footnote 10.
[49] *Defeating Darwinism,* 15.
[50] *Defeating Darwinism,* 17.
[51] *Defeating Darwinism,* 17.

these individuals deserve to be called deists, for their God is not the God of the Bible. In contrast, many Christian biologists and I acknowledge the Creator's wisdom and power in the physical laws and processes (such as evolutionary processes) that he has instituted — *as well as* his direct activity in our lives, including dramatic occurrences of his grace in charismatic experiences. *Mistake number two* for Johnson is this: It is *not logically necessary* to claim that if God used an evolutionary process in creating his wondrous universe (including us) then he must have retired outside it, is not active in it, or does not care for it.

Finally, Emilio's "mistake number three," Johnson asserts, is "the idea that religious statements belong to the realm of faith while scientific statements to the realm of reason."[52] "This rational defensive strategy born of desperation," Johnson claims is "the error that many prominent theologians and philosophers have made," that "gives away the realm of reason."[53] He goes so far as to suggest that this so-called "retreat to the faith escape" is exactly what non-Christians hope Christians will affirm. In order to challenge Johnson's thinking here, let us take for the sake of illustration an example very dear to the hearts of all Christians including both Johnson and me — the precious blood of Jesus. If time machines existed, then we might return to the first century to analyze our Lord's blood with the many analytical tools we have in medicine. Upholding that Jesus was indeed fully a man as the Church creeds through the ages have attested (and as Johnson and I believe), we could determine his red and white blood cell count, his Rh-factor, and blood type (A, B, O), and even whether he had antibodies to certain viruses like hepatitis B. In other words, we could *know* Jesus' blood through *scientific knowledge*. However, the analytical tools of science will *never ever* be able to record the cleansing power of the blood of God's Lamb in our hearts. Similarly, there are no scientific sensors to record my sinfulness. But is the power of Jesus' blood real? Absolutely! I have experienced it in my life, as has Johnson and the multitudes of sinners who have come face-to-face with the living Christ. As a result, we come to *know* this reality through *religious knowledge*. Johnson's *mistake number three* is as follows: he fails to grasp fully the relationship between reason and faith, and that these are different realms of knowing and experience. This does not mean that one is better than the other. Rather, both are valid ways of knowing, but dif-

---

[52] *Defeating Darwinism*, 19.
[53] *Defeating Darwinism*, 17, 20.

ferent ways of knowing, and together they complement each other, giving us a complete worldview as God has intended.

Johnson's use of Emilio is a classic example of the straw-man argument (more about this in the next section). It is effective because in the first chapter of *Defeating Darwinism* he has his readers believe that: (1) 'evolution' means unguided, purposeless change, (2) evolution at best can have only a deistic god, and (3) Christians sell out on the faith if they acknowledge that there is a distinction between religious and scientific statements. Put together in this fashion, and if accepted by the reader, Johnson then can conclude that Emilio and Christian evolutionists "mistakenly think they have resolved the problem by viewing evolution as a God-guided system of gradual creation."[54] I do not believe Emilio would agree with the way Johnson has outlined his position. As noted, there is a valid argument against each of these three points. But more importantly, by setting up his argument in this fashion early in *Defeating Darwinism*, Johnson quickly closes the door on any serious consideration that Jesus created our world through an evolutionary process.

Turning the Baloney Detector on *Darwinism Defeated*

Johnson dedicates an entire chapter to the "baloney detector" which he defines as "simply a good grasp of logical reasoning and investigative procedure."[55] He suggests that we point the detector at "claims made by the official scientific establishment," and particularly at the theory of biological evolution. By understanding the "varieties of baloney" (e.g., straw-man and *ad hominem* arguments), Johnson maintains that the reader will gain the skills to become a good critical thinker. But let us also point the detector at *Defeating Darwinism* to see if we can detect any of these "varieties" in his book.

Straw-man Arguments

Johnson correctly states that "a 'straw man' argument distorts somebody's position in order to make it easier to attack."[56] But consider how Johnson describes the views of his Christian friend Emilio in the first chapter of his book:

> In that case the scientific facts do not matter—whatever they are. Emilio
> can proclaim with relish a complete naturalism in science and insist that

---

[54] *Defeating Darwinism*, 22.

[55] *Defeating Darwinism*, 37.

[56] *Defeating Darwinism*, 41.

it makes no difference to faith. He might explain this position in words like these: 'Yes, the diversity and complexity of life are the result of evolution. Yes, evolution is a blind, unsupervised and unintelligent process. Yes, we humans are the result of a purposeless and natural process that did not have us in mind.'[57]

As noted earlier, we do not know who Emilio is because Johnson has changed his name. However, *Defeating Darwinism* includes a short e-mail message written by this Christian student and we can glean some of his theology from it.[58] It is clear that Emilio believes: (1) he has been a creationist all his life, (2) the first chapter of Genesis is not to be interpreted literally; (3) God is the transcendent Creator of both time and space; and not bound by either; (4) God is the Creator of all the physical laws; (5) God intentionally created life through an evolutionary process; (6) evolutionary biology is the science that investigates how God created life; and (7) well-intentioned Christian antievolutionists are a stumbling block to the cause of Christ, indicating Emilio's pastoral concern for unbelieving colleagues. In sum, Emilio's view of evolution is *not* that of "an unsupervised, impersonal, unpredictable and natural process" as Johnson depicts. It might be added that the scientific facts do matter for Emilio because he had the *integrity* and *open-mindedness* to view the primary evidence first-hand by pursuing biology at university-level. Furthermore,

---

[57] *Defeating Darwinism*, 18.

[58] Here is Emilio's entire e-mail to let readers judge for themseves Johnson's interpretation of this student's theology:

"I've been a Christian and a creationist all my life, fiercely against evolutionism until I started my biology course at the university and began learning about evolution. Guess what: I am still a creationist and now I am also an evolutionist! It has become clear to me that the first chapter of Genesis is an allegory (if not check how it states that there was morning and evening before the creation of the Sun and Moon—an impossibility), and once we accept this there is no reason why God could not have created all there is in as many million years as you wish.

If God created time and space, then he is outside of it and therefore is not affected by it—time has no meaning to God! I believe that God created the laws of physics, and therefore everything that results from such laws is God's creation. To say that the species evolved does not deny God's act of creation. Quite the opposite: evolution is the science that studies how God created the species.

Furthermore, evolutionism and creationism cannot be put in the same category, as one is science, of the rational, and the other is faith, of the supernatural.

I am a Christian, and it offends me to see that Christians are being viewed as lamebrains just because some well-intentioned but ignorant brothers of mine try to discuss such matters without scientific knowledge—please stop, you do more harm than good." *Defeating Darwinism*, 13–14.

he had the *courage* to challenge what seems to be his fundamentalist upbringing which featured young earth creationism, and then to *reformulate* his views on origins and his interpretation of Genesis. I am disappointed that Johnson has misrepresented Emilio's views, and I believe Johnson's "baloney detector" would register this as a straw-man argument.

Johnson also makes a straw man out of Charles Darwin. After reading any of Johnson's three books one cannot help but get the impression that the nineteenth-century biologist insisted that evolution was dysteleological. But as noted earlier, this is historically inaccurate. It is true that Darwin certainly was not a Christian, yet there is ample evidence from both his public and personal writings to suggest that, though he had agnostic moments, he believed in a Creator. That is, Darwin was not the *naturalist* or *materialist* Johnson claims to have "put on trial." Instead, if Johnson would approach the primary literature on Darwin with historical sensitivity, he could make a strong case for the powerful impact that design in nature has on the human psyche. Comments on design made by Darwin throughout his life and up until the year of his death testify to the profound reality that "The heavens declare . . ."

The straw-man argument is also seen in Johnson's misrepresentation of the modern theory of evolution. He consistently depicts Darwin's nineteenth-century theory as the modern position, thus making it easier to attack. To be sure, Darwin's work is a milestone in biological thought, but the theory of evolution has moved on quite a distance from when he first formulated it. Attacking errors that Darwin made nearly 150 years ago makes about as much sense as pointing out some of Galileo's scientific oversights in order to cast suspicion on modern astronomical theory.[59] More specifically, Johnson insists that gradualism is "essential to Darwinism."[60] This was indeed a foundational principle in Darwin's theory. But with the advance of science it has become clear that species did not evolve in the way Darwin had conceived, and modern evolutionary theorists (and not antievolutionists) have modified the theoretical model in the light of new discoveries (thus pointing to the integrity of the scientific community and its willingness to dismiss even Darwin's most important ideas). It is interesting that Johnson used Stahl's textbook *Verte-*

---

[59] For example, Galileo believed that the sun was literally the centre of the universe and that a literal firmament of the "stellar sphere" surrounded the planetary orbs. See Galileo's letter to Grand Duchess Christina (1615) in Maurice A. Finocchiaro, *The Galileo Affair: A Documentary History* (Berkeley: California University Press, 1989): 118–119.

[60] Johnson's emphasis that gradualism is "essential to Darwinism" is best seen in *Darwin on Trial*, 38, 41, 50–2, 67, 137–7, 144, 179, 183; in particular on 33 and 43.

brate History: Problems in Evolution (1985) as his "primary source" of information about the vertebrate fossil record. In the preface to the edition that Johnson consulted, she states that the problem with the modern antievolutionists is that they take

> advantage of the popular interest in the scientific debate between the advocates of Darwin's gradualism and a punctuated equilibrium . . . pointing to the disagreement among scientists about the evolutionary process as evidence for the error of the whole Darwinian theory.[61]

This is the very aspect of the scientific debate that Johnson exploits throughout his critique of evolution. With Darwin's gradualism "defeated," Johnson's next move is to assert, "It seems for now as if new forms appeared mysteriously and by no known mechanism at various widely separated times in the earth's history."[62] Of course, for Johnson the 'mystery' is solved with God's direct intervention. However, what happens if science comes to understand fully the mechanisms for the rapid evolution of species as depicted in the fossil record?[63] More importantly, what happens to Johnson's view of God? Are we willing to extrapolate from our ignorance of the mechanisms today that no such mechanisms exist to account for evolution? And yes, maybe even too bluntly, is the evangelical church willing to base a belief that there are no such mechanisms on the *judgment* of a non-biologist? Should such a judgment not be made by a committee of professional biologists who are trained in areas of biology that are relevant to this very topic? Please note, this is not a personal attack against Johnson. Rather these are reasonable and respectful questions that must be asked before evangelicals begin to declare where God has directly intervened in the history of life, and what exactly he did.

Ad Hominem *Arguments*

Johnson correctly defines *ad hominem* arguments as those that "attack the person making the argument instead of the argument itself."[64] Dis-

---

[61] Stahl, *Vertebrate History*, ix. Punctuated equilibrium is the view that species evolve rapidly during short periods of time while they remain morphologically the same during extended periods. As a result, fewer transitory forms are expected in the fossil record.

[62] *Defeating Darwinism*, 67.

[63] The discovery of mechanisms for rapid evolution is an important part of the new biological subdiscipline of evolutionary developmental biology. This study shows great promise in finding how small changes in the genetic code of major control genes could result in large and dramatic morphological change in living forms.

[64] *Defeating Darwinism*, 40.

appointingly, Johnson fails to adhere to his own admonition against this way of arguing. He often questions the intellectual ability of scientists, and even goes so far as to raise suspicion of their personal integrity. And this method of argument is even launched at fellow believers—Christian theologians. Here are but a few examples taken from *Defeating Darwinism* to let readers judge for themselves.

Regarding scientists, Johnson asserts:

> When someone claims to have magical powers, the claims must be tested before an audience of stage magicians, who know how the tricks of illusion are done. *Scientists are notoriously easy to fool in such matters.* When dealing with an ideology like Darwinism, the critical audience needs to include professors of rhetoric and legal scholars, who are skilled at spotting question-begging assumptions and similar *tricks* of logic.[65]

> My experience speaking and debating on this topic at universities has taught me that scientists, and professors in general, are *often confused* about evolution. They may know a lot of details, but they *don't understand the basics.* . . . If high-schoolers need a good high-school education in how to think about evolution, professors and *senior scientists* seem to need it just as badly.[66]

> Think how easy it would be for ambitious fossil hunters to *fool* themselves, when the reward for doing so may be a cover story in the National Geographic and a lifetime of research funding. Think how much pressure the other physical anthropologists are under to develop standards that will allow *some* [Johnson's italics] fossils to be authenticated as human ancestors. A fossil field without fossils is a candidate for extinction.[67]

> Evolutionary biologists have been able *to pretend to know* how complex biological systems originated only because they treated them as black boxes. Now that biochemists have opened their black boxes and seen what is inside, they know the Darwinian theory is *just a story*, not a scientific explanation.[68]

To summarize, Johnson believes that scientists are notoriously easy to fool, confused, pretenders, storytellers, pressured by pride and profes-

---

[65] *Defeating Darwinism*, 115. My italics.
[66] *Defeating Darwinism*, 11. My italics.
[67] *Defeating Darwinism*, 62. My italics.
[68] *Defeating Darwinism*, 77. My italics.

sional security, and in dire need of a high-school level education on evolution even for the senior members. Note also Johnson's own self-endorsement into the origins debate as a professor of rhetoric and legal scholar in the first passage. But is it true that scientists are such a sorry intellectual group? For Johnson's assessment of modern scientists to be accurate, he has to postulate some sort of double-mindedness in them. On the one hand, he openly accepts an old universe, affirming the Big Bang theory of the cosmologists and the earth's developmental history as given by the geologists.[69] Moreover, I am sure Johnson must marvel at other scientific achievements such as landing men on the moon or a rover on Mars and the life-saving techniques of modern medicine. Yet despite these incredible achievements, it seems that with regard to the topic of evolution there is a complete and thorough breakdown of the scientists' ability to think and act with integrity. But is such a double-mindedness indeed the case with the modern scientists? How does Johnson account for this? Will he argue that the minds of scientists are clouded by a spiritual darkness when dealing with biological origins . . . yet these same minds are open and acute with regard to cosmology and geology? Or inversely, is this *perceived* double-mindedness symptomatic of the fact that Johnson and the North American evangelical church have yet to deal directly and fully with biological evolution? As a body of believers we have come to terms with cosmology and geology in the last twenty or so years, but is our reticence toward the theory of evolution simply because we have avoided it?

Turning to Christian theologians, Johnson claims:

> Darwinian evolution is a theory about how nature might have done this [i.e., created life] without assistance from a supernatural Creator. That is why 'evolution' in the Darwinian sense is by definition mindless and godless. *Pretending* otherwise is an *evasion* of the conflict, not a resolution of it. Yet many Christian theologians and educators take this *evasive* approach because they are hoping to find an easy way *to avoid coming to grips* with a very difficult problem. [70]

---

[69] Considering that Johnson accepts the standard cosmological model, it must be asked whether he claims that there were divine interventions between the time of the Big Bang and the beginning of life on earth (a period of roughly 10 to 15 billion years). Using his argument that a "Creator who merely sets a process in motion and thereafter keeps hands off is easily ignored," (Johnson, *Reason in the Balance*, 77) one would expect a number of supernatural interventions during this long span of time.

[70] *Defeating Darwinism*, 15–16. My italics.

> Even Christian college and seminary professors are bound to be influenced by the spirit of the times. To be *successful* in academic life is to be current with the *fashionable* thought from the most prestigious universities, and teachers *can hardly help absorbing* the ways of thinking that they themselves have been taught.[71]

According to Johnson, it seems that Christian theologians share with modern scientists in being "pretenders" led by professional success and intellectual fashion. In addition, on one hand they do not seem to have the ability to transcend the intellectual milieu within which they exist and are caught in the "spirit of the times;" yet on the other hand, they also seem to appreciate it enough to take "evasive" actions to avoid conflict. Assuming that Johnson is using the terms evasive and evasion accurately, he is issuing quite an indictment against fellow brothers and sisters in Christ who labour in academic theology; for these terms are defined as "not direct, candid, or forthright," "shifty," and "dodging or circumventing a law, responsibility, or obligation."[72] And so, as with the modern scientists, for Johnson's assessment of Christian theologians to be true, it appears that he also has to postulate their double-mindedness (or should we say double-spiritedness?). On the one hand, they can affirm the reality of the Christ of the New Covenant, yet on the other hand in regard to the origins debate they are "evasive" and led by the "spirit of the times." But is this indeed the case for any theologian opposing Johnson's views?

Disappointingly, Johnson's use of *ad hominem* arguments against scientists and Christian theologians is not limited to *Defeating Darwinism* since this approach is found in his other two books — *Darwin on Trial* and *Reason in the Balance*. Most will agree that this method of arguing only discourages open dialogue.

Trustworthy Experts and Appeals to Authority

Johnson is correct in claiming that we are often forced to trust experts because "we can't possibly think out everything for ourselves all the time."[73] This is particularly true in attempting to understand a topic as multidisciplinary as the origin of life. Johnson offers some valuable advice in dealing with this problem:

---

[71] *Defeating Darwinism*, 91. My italics.

[72] "Evasion" and "evasive" in *Webster's Third International Dictionary* (Springfield, Mass.: Merriam-Webster, 1986): 787.

[73] *Defeating Darwinism*, 45.

Trustworthy experts are ones who *understand their responsibility* to give us their expertise *without claiming to know more than they really do*. Really trustworthy experts don't try to evade our baloney detectors, and even warn us to watch out for their own expert bias.[74]

In the light of his judicious suggestion, I believe it would be fair to ask a couple of questions to Johnson: (1) Regarding the primary evidence for evolution, which he deems deficient, is it possible that he is claiming to know more about the fossil record than he really does? (2) Where in his writings does he explicitly warn us to watch out for his own expert bias? In considering the first question, it is Johnson's experience that

evolutionists often proudly point to a *small number* of fossil finds that *supposedly confirm* the theory. These include the venerable bird/reptile *Archeopteryx*, the "whale with feet" called *Ambulocetus*, the therapsids that *supposedly link* reptiles to mammals, and especially hominids or ape-men, like the famous Lucy. . . . I am not as impressed by such examples as Darwinists think I should be, because I know that the fossil record overall is extremely disappointing to Darwinist expectations.[75]

Johnson gives the impression he is familiar with the fossil record and that it is quite deficient. However, consider his criticism regarding the extinct order of therapsids that "supposedly link" reptiles and mammals. Johnson deals with this fossil evidence for evolution in less than three pages in *Darwin on Trial* and judges that this data is at best superficial.[76] However, Johnson fails to appreciate the large number of therapsids that have been collected and catalogued by paleontologists, nor does he seem to recognize the vast literature written on this subject. Just a brief scan at the appendix listing therapsids in Carroll's textbook *Vertebrate Paleontology* (1988) will show nearly 250 species of this ancient order (there are literally thousands of specimens of therapsids in museums). This is *not* an insignificant and "small number of fossil finds that supposedly confirm the theory" of evolution.

The situation appears to be the same as with Johnson's knowledge about whale evolution in that there exists more fossil evidence than he seems to be aware of. Interestingly, his mention of "the 'whale with feet' *Ambulocetus*" in *Defeating Darwinism* demonstrates the *fruitfulness* of the theory of biological evolution. We will recall that in *Darwin on Trial* Johnson was quite suspicious of the idea that the limbs of the purported an-

---

[74] *Defeating Darwinism*, 46. My italics.
[75] *Defeating Darwinism*, 59–60. My italics.
[76] *Darwin on Trial*, 77–80.

cestors of whales could have been reduced to flippers, and he even had grave doubts of the interpretation that the reduced legs found in association with one fossil specimen of *Basilosaurus* actually belonged to this whale. However, Johnson is now aware of evidence for the evolution of limbs into flippers, mentioning the discovery in 1994 of *Ambulocetus* with reduced front legs and toes terminated by a convex hoof (recall that the theory of whale evolution suggested that whales descended from hoofed animals—mesonychids).[77] This is a positive change considering that after discussing whale evolution in both editions of *Darwin on Trial* (1991 and 1993) Johnson complained that "to Darwinists unsolvable problems are not important," and that "the fossil evidence was heavily against [Darwin's] theory, and this remains the case today." But the theory of whale evolution predicted that over time more fossil evidence would emerge to strengthen the hypothesis—and it did with the discovery of this ancient whale. It would appear that the gap in the whale fossil record that Johnson proclaimed in 1991 and 1993 has been filled in part by "the 'whale with feet' *Ambulocetus*." It would be interesting to see what Johnson's views would be regarding whale evolution if he were familiar with all the primary fossil data and scientific literature rather than limiting the evidence to Stahl's dated textbook and news reports of recent fossil findings.

In considering the second question in this section, where in *Defeating Darwinism* does Johnson explicitly warn us to watch out for his own expert bias? The answer is that he never does. Instead, he appeals to his academic authority, claiming that the origins debate requires a "critical audience" that includes "professors of rhetoric and legal scholars, who are skilled at spotting question-begging assumptions and similar tricks of logic."[78] On the surface, such an "expert bias" is clearly attractive to everyone concerned with the pursuit of truth. But the reality is that Johnson has an important *theological* bias that shapes his view of origins. More specifically, two theological concepts—*interventionism* and *concordism*[79]—

---

[77] Thewissen et al. "Archaeocete Whales," 211.

[78] *Defeating Darwinism*, 115.

[79] It is important to define these two terms. First, in discussions on the topic of origins, the term *interventionism* is used to describe God's direct activity in the creation of the universe and (in particular) life. Evolutionary creationists would suggest that this type of 'interventionism' was not employed by God when he created. However, they openly affirm that the Bible's record of divine activity with the children of Israel and the church is clearly interventionistic. As a result, there are two types of interventionism: (1) *cosmological interventionism* in the creation of life, and (2) *personal interventionism* in the lives of individuals and communities. It is possible to view God's action in the creation of life as non-

operate at a deep level in his thinking and only occasionally do these openly emerge in his writings.

Johnson maintains that a "Creator who merely sets a process in motion and thereafter keeps hands off is easily ignored."[80] According to Johnson the only God worthy of praise and worship would be one who actively intervenes in his creation throughout time. This God-of-the-gaps view is one of the primary *assumptions* underlying his arguments. As a result, it is understandable why Johnson enthusiastically endorses Michael Behe's *Darwin's Black Box* (1996). Behe's notion of the purported *irreducible complexity* of biomolecular structures calls for divine intervention to account for their existence since he claims that known natural processes could not account for their emergence.[81] Similarly, with the

---

interventionistic and his activity with his image-bearing creatures as interventionistic. This would be my view. Phillip Johnson does not distinguish between these two types of interventionism, and my claiming that he is an 'interventionist' must be understood in in the context of the origins debate in that he asserts that God intervened directly in the creation of life (i.e., cosmological interventionism).

Second, the term *concordism* as it is employed in the debate on origins deals with whether there are scientific truths in Genesis 1–11 that relate or correspond to modern science. However, concordism extends well pass the origins debate and three forms of concordism must be distinguished: (1) *cosmological concordism* suggests the existence of a correspondence between the origin of the universe and life and the early chapters of Genesis 1–11; (2) *historical concordism* refers to the fact that the Bible is indeed a historically reliable document as testified by the academic specialty of biblical archeology (e.g., king lists, nation states, cultures, battles, etc.); (3) *spiritual concordism* is the most important form of concordism. It suggests that there is a significant correspondence between the theological truths in the Bible and the spiritual realities we experience in our lives. For example, a simple reading of the Garden of Eden account will powerfully convict the sincerely seeking individual of his/her sinful nature. Again, my calling Johnson a "concordist" must be understood in the context of the origins debate. He maintains that Genesis 1–11 contains cosmological truths that correspond with modern science (i.e., cosmological concordism). In contrast, an evolutionary creationist would argue that these early chapters of the Bible only offer a spiritual concordism (e.g., the universe is a creation of a Creator, the creation is good, humanity bears the image of God, sin is utterly real and utterly significant).

[80] *Reason in the Balance*, 77.

[81] Johnson claims, "There is therefore no pathway of functional intermediate stages by which a Darwinian process could build such a system [i.e., molecular structures like cilia] step by tiny step" (*Defeating Darwinism*, 77). It must be underlined that the theory of evolution was built on the fossil record of *tissues* and not *biomolecular structures* like Behe's often used example of cilia. The reason for this is that biomolecular structures like cilia are too small to be fossilized. As a result, Behe and biochemists are at a disadvantage in attempting to reconstruct evolutionary precusors in that they simply do not have a biomolecular fossil record from which to work. In contrast, tissues like bones and teeth can be easily fossilized as the geological record shows. Let me offer an illustration from my own biological specialty—the evolution of teeth and jaws—to help further explain the problem

Cambrian explosion (the 'biological Big Bang') Johnson emphasizes the speed at which organisms appeared, for he suggests this implies God's direct action. Problems with gaps in the fossil record or the scientific community's openly attested difficulties in determining prebiological evolution are also deemed confirmatory evidence, according to Johnson, for God intervening in gaps.

The second theological concept that shapes Johnson's view of origins is concordism. That is, he suggests that there are scientific truths within the first chapters of the Bible that have been confirmed by modern science. He writes:

> Evolution within species is as much a biblical doctrine as a scientific one, for the Bible taught us (long before modern science) that all different races of man descend from a common human ancestor. Finch-beak variation in no way denies that only God can make a bird.[82]

To be sure, over the years Johnson has been very cautious in trying not to bring the Word of God into the origins debate, but this is one of the rare occasions where it is clear that theological categories are a factor in his science. As discussed earlier, progressive creationism suggests that there is a certain biological plasticity that allows for some evolutionary change such as in finch-beak variation. Yet this change is confined within certain parameters, for example, within a species as Johnson suggests in this passage.

Johnson's view of divine interventionism in creation and his biblical concordism make him a progressive creationist. But nowhere in his writings does he warn us to watch out for this expert bias. Nor does Johnson attempt to justify these theological assumptions to his readers. But as history records, the God-of-the-gaps approach is vulnerable to the advance of science and has led to disastrous pastoral consequences. The closure of alleged gaps results in the retreat of this type of God into narrower recesses of human ignorance. With regard to concordism, it is clear that Johnson does not appreciate the nature of the opening chapters of God's precious Word. Again Church history has revealed the consistent

---

Behe has in entering the origins debate. If I were given only a modern human dentition, and I was not aware of the evolution of teeth and jaws as seen in the fossil record, then I cannot imagine how I could ever come up with the series of dental precusors known from the fossils. This is the very problem Behe and his colleagues have as biochemists in the origins debate. They only have the modern form of cilia and no fossil evidence of cilia to aid in determining the series of stages that this biomolecular structure passed through in its evolution.

[82] *Defeating Darwinism*, 87.

failure of concordist interpretations of the Scriptures.[83] Most Christian believers would agree that we should not use the Bible in constructing scientific theories on astromony (e.g., an earth-centred universe) or reproductive biology (e.g., the idea that infertility is limited to barren women). The use of Genesis 1 to justify a view of biological origins is every bit as precarious.[84]

Finally, one last issue with regard to the topic of the trustworthiness of experts that is not found in *Defeating Darwinism*, but that I believe is worth serious consideration. It must be reported that Johnson also applies his "specialty in analyzing the logic of arguments and identifying the assumptions that lie behind those arguments" in the world of clinical medicine. Interestingly, he is a member of the Group for the Scientific Reappraisal of the HIV/AIDS Hypothesis. In a 1995 letter to *Science*, the prestigious journal of the American Association for the Advancement of Science, Johnson and his colleagues write:

> In 1991, we, the Group for the Scientific Reappraisal of the HIV/AIDS Hypothesis, became dissatisfied with the state of the evidence that the human immunodeficiency virus (HIV) did in fact cause AIDS. . . . [T]he correlation of HIV with AIDS, upon which the case for HIV causation rests, is itself an artifact of the definition of AIDS. . . . The bottom line is this: [we] the skeptics are eager to see the results of independent scientific testing. Those who uphold the HIV "party line" have so far refused. We object.[85]

Johnson is very correct in stating that "trustworthy experts are ones who *understand their responsibility* to give us their expertise *without claiming to know more than they really do.*"[86] However, one can only hope that if

---

[83] For a excellent review of the many failed concordist attempts, see Stanley L. Jaki, *Genesis 1 Through the Ages* (London: Thomas Moore, 1992).

[84] Of course, determining the nature or literary genre of Genesis 1 is a study in itself. Suffice to say that though we evangelicals have a wonderfully 'high' view of the Scriptures as being the Word of God, inspired by the Holy Spirit, we have yet to grasp a principle elaborated by one of our theologians George Eldon Ladd: "the Bible is the Word of God given in the words of men in history." George Eldon Ladd, *The New Testament and Criticism* (Grand Rapids: Eerdmans, 1967): 12. For an excellent book that grasps the historicity of the Word of God as it relates specifically to the origins debate see Paul H. Seely, *Inerrant Wisdom: Science and Inerrancy in Biblical Perspective* (Portland: Evangelical Reform, 1989).

[85] Eleen Baumann, Tom Bethell, Harvey Bialy, Peter H. Duesberg, Ceila Farber, Charles L. Geshekter, Phillip E. Johnson, Robert W. Maver, Russell Schoch, Gordon T. Stewart, Richard C. Strohman and Charles A. Thomas, Jr., "AIDS Proposal," *Science* 267 (17 February 1995): 945–946. Also see Phillip E. Johnson, "AIDS and the Dog that Didn't Bark," *Insight* (14 February 1994): 24–26.

[86] *Defeating Darwinism*, 46. My italics.

Johnson is going to make public statements about the HIV/AIDS hypothesis then he should become very familiar with the primary medical literature on the subject.

The Conflation Factor Again

In pointing Johnson's Baloney Detector at *Darwinism Defeated* at least three varieties of baloney were noted. It is clear that Johnson employs straw-man and *ad hominem* arguments, and that he overstates his authority as a trustworthy expert. To these it must be added that the problem of conflation is also evident in this book. As noted earlier, Johnson's antievolutionism in his first two books gained authority by conflating it with two powerful truths—the reality of design in nature and the philosophical vacuity of materialism/naturalism. This rhetorical move is also made in *Darwinism Defeated*. Conversely, a powerful truth can be discredited if juxtaposed to erroneous notions or to ideas that arouse strong negative emotions. Johnson uses this second approach in the closing chapters of *Darwinism Defeated* by listing numerous sinful situations that elicit a powerful negative response in any committed Christian: the increase in the church of liberal ministers and theologians (e.g., John Shelby Spong) and their 'demythologizing' of the Scriptures, claiming that the biblical writers were offering mere flights of fancy; the consequences of the sexual revolution including increased rates of divorce, illegitimacy, premarital sex, and homosexuality; radical feminism and the right to abortion; and the School Prayer Decision in 1962 that banned prayer in public school. The underlying message, of course, is that these are all the sour fruits of an evolutionary view of origins. The association of these sinful situations with the theory of evolution ultimately elevating and 'authorizes' Johnson's antievolutionism in the eyes of his evangelical readers.[1]

Are Evangelicals Inheriting the Wind?

Before attempting to answer the question in the title of this section, I believe it is necessary to deal with what I deem to be the most important issue for Christians in the origins debate—the pastoral implications.

Pastoral Implications of the Origins Debate

The topic of origins is indeed important for theology, but not so important as to become the central issue of faith. As others before have admonished, "Don't get the Rock of Ages mixed up with the age of the rocks." It is because our faith rests upon Jesus Christ that we are *Christians* first and foremost. Three important relationships can be affected by our views of origins: (1) our relationship with other Christians, (2) our relationship with our children, and (3) our relationship with non-Christians.

First, how do Christians with different views of origins relate to one another? Is one's orthodoxy and love for the Lord determined by how one conceives God's method of creation? Is this issue big enough to cause *division* among Christians? Or is it a *difference* among Christians that we should be able to live with in the church? (See 1 Cor. 11:18–19) Personally, I would be the first to pass the communion cup to my young earth creationist or progressive creationist brothers and sisters. Unfortunately, Johnson's open and direct attack on Christian theologians and educators inflames an already tense situation in the church, and this approach will only be an alienating factor among members of the body of Christ.

Second, what are we going to teach our children regarding origins? Imagine just for a moment that the Lord did indeed use an evolutionary process in creating the universe and life. What happens to the child who is taught young earth creationism or progressive creationism in a Christian school or a church Sunday school, and then sees the scientific data for evolution first-hand in the university or paleontological museum? Those who have seen such a scenario unfold often report disastrous spiritual consequences. Considering that Johnson's *Defeating Darwinism* was written "for late teens — high-school juniors and seniors and beginning college undergraduates, along with the parents and teachers of such young people,"[87] it is clear what he wants our children to believe. And there are at least three good reasons why he may well be successful. First, North American evangelicalism has yet to come to terms with biological evolution. Antievolutionism operates almost as an unwritten article of faith in our churches. Second, Johnson's meteoric rise to leadership in North American evangelicalism places him as the most influential spokesman on the topic of origins, and many will accept his views. Lastly, the home-schooling movement has gained wide acceptance in evangelical circles. The need for a readable critique of evolution could see Johnson's *Defeating Darwinism* easily incorporated into these programs. But the most foundational question we must ask ourselves for the sake of our children's education is this: Considering the importance of

---

[87] *Defeating Darwinism*, 9–10.

this issue, is it judicious, yes, even logical to permit someone with no formal training in biology to determine what should be taught regarding a biological topic such as evolution?

Finally, what are we going to tell the unbelieving world about the origin of life? Upholding the importance of proclaiming the Good News, I am very sensitive to Paul's admonition that we Christians should not be a stumbling block to the watching world (2 Cor. 6:3). Again, assume that the Lord did indeed use an evolutionary process in creating the universe and life. Can you imagine how much of a stumbling block young earth creationism or Johnson's progressive creationism could be to those who see the scientific data for evolution daily? Many Christians, like Johnson, weld their antievolutionism to the Cross of Christ and their faith. But to my chagrin, I have too often seen in the university environment that such a conflation has non-believers disregard the Cross as they angrily mock scientific understandings of the antievolutionists. History reveals that this problem is not new, as St. Augustine commented many years ago:

> Usually, even a non-Christian knows something about the earth, the heavens, and the other elements of this world, about the motion and orbit of the stars and even their size and relative positions, about the predictable eclipses of the sun and moon, the cycles of the years and seasons, about the kinds of animals, shrubs, stones, and so forth, and this knowledge he holds to as being certain from reason and experience. Now, it is a disgraceful and dangerous thing for an infidel to hear a Christian, presumably giving the meaning of Holy Scripture, talking nonsense on these topics; and we should take all means to prevent such an embarrassing situation, in which people show up vast ignorance in a Christian and laugh it to scorn. The shame is not so much that an ignorant individual is derided, but that people outside the household of the faith think our sacred writers held such opinions, and, to the great loss of those for whose salvation we toil, the writers of our Scripture are criticized and rejected as unlearned men. If they find a Christian mistaken in a field which they themselves know well and hear him maintaining his foolish opinions about our book, how are they going to believe those books in matters concerning the resurrection of the dead, the hope of eternal life, and the kingdom of heaven, when they think their pages are full of falsehoods on facts which they themselves have learnt from experience and the light of reason?[88]

---

[88] St. Augustine, *The Literal Meaning of Genesis*, 2 vols., trans. J. H. Taylor (New York: Newman, 1982): 42–43.

My prayer and hope is that every Christian entering the origins debate take St. Augustine's admonition seriously. In particular, that a rigorous and judicious self-examination of one's knowledge and the relevance of that knowledge to the problem of origins be done within the context of a community familiar with the issues. At the bare minimum, please do not conflate the Cross of our Lord and Saviour with any view of origins.

The Wedge:
Another Example of the Scandal of the Evangelical Mind?

The first sentence of Mark Noll's 1994 book *The Scandal of the Evangelical Mind* is a blunt indictment: "The scandal of the evangelical mind is that there is not much of an evangelical mind."[89] However, it must be understood that Noll is a thoroughly committed evangelical who is a professor at a leading evangelical college, and he admits writing as "a wounded lover." More specifically, Noll loves the life of the mind, but in doing so he has been wounded by the anti-intellectualism of his theological tradition. Though his review of modern evangelical intellectual history is critical, it is also intended to be constructive, and ultimately his work is a *cri de coeur*. George Marsden accurately summarizes Noll's admonition to the evangelical Church, "If Christians are to serve God with their minds, they must do their homework."[90]

Johnson has a plan to challenge and correct what he deems science's ideological commitment to materialist/naturalist philosophy, and he calls his strategy "the wedge." He claims that his "own books (including this one [i.e., *Defeating Darwinism*]) represent the sharp edge of the wedge" that will split the "log" of evolutionary theory and materialism/naturalism.[91] Johnson adds that Michael Behe's *Darwin's Black Box* is the "first broadening of the initial crack" that he opened. Further, he boasts, "I find as time goes by that my greatest satisfaction comes not from the work I can do myself but from the accomplishments of younger people to whom I have given encouragement and for whom I have opened doors."[92] And concluding, Johnson predicts that evolutionary

---

[89] Mark A. Noll, *The Scandal of the Evangelical Mind* (Leicester, England: Inter-Varsity Press, 1994): 3.

[90] George M. Marsden, quoted from the jacket of *The Scandal of the Evangelical Mind*.

[91] *Defeating Darwinism*, 92.

[92] *Defeating Darwinism*, 96.

theory and materialism/naturalism "will collapse with astonishing swiftness."[93]

However, I believe that Johnson has not done his homework on the theory of biological evolution. To be sure, *Defeating Darwinism* will be well received in the evangelical community—but only because this body of believers in Christ has yet to come to terms with evolution. And to my chagrin, *Defeating Darwinism* will probably become part of the home-school curriculum in evangelical households. However, in the spirit of St. Augustine, if we evangelicals are going to deal with the issue of biological origins with integrity, it behooves us to enter through the doors of the biological academy and to do our homework.

I respectfully submit that the Phillip E. Johnson phenomenon is another incident in the history of evangelicalism that supports Mark Noll's intellectual scandal thesis. As "a wounded lover" because of my acceptance of evolutionary creationism, my *cri de coeur* is to proclaim that the only "wedge" Johnson is introducing into our society is a wedge between the evangelical church and the modern university. *To conclude, the current popularity of Professor Johnson's antievolutionism in North American evangelicalism is a clear example of this Christian community inheriting the wind.*

Finally, I will end this paper with a question directed specifically to my brothers and sisters in the North American evangelical church. Currently, the three leading antievolutionists who shape the beliefs of our community are Henry Morris, Hugh Ross, and Phillip Johnson. Each holds a respected earned doctorate (engineering, astronomy, and law, respectively), but in a discipline that is not relevant to evolutionary biology. I am certain that if they or any evangelical reading this paper were to have a decayed tooth, they would surely make their way to the office of a dentist in good standing with his/her professional association. Clearly, no one would seek the services of an engineer, an astronomer, or a lawyer to deal with this problem. My question then is this: If evangelicals will not open their mouths to anyone less than a licensed dentist to get a tooth filled, why then on the important topic of biological origins do evangelicals open their minds and hearts to be filled with the speculations of non-biologists like an engineer, an astronomer, and a lawyer? Our children deserve better.

---

[93] *Defeating Darwinism*, 114.

Dedication

To Dr. Keith Miller who courageously stood up to defend science at the 1997 meeting of the American Scientific Affiliation in Santa Barbara, California, and who later quietly wept for this association of believing scientists.

Postscript:
Revision of the National Association of Biology Teachers' Statement on Teaching Evolution

On the day I was to send this paper to the publisher, an encouraging *preliminary* report was released regarding the revision of the 1995 "Statement on Teaching Evolution" by the National Association of Biology Teachers.[94] The primary tenet of the 1995 statement states:

> The diversity of life on earth is the outcome of evolution: an unsupervised, impersonal, unpredictable and natural process of temporal descent with genetic modification that is affected by natural selection, chance, historical contingencies and changing environments.[95]

As I commented in my paper, Johnson was correct in pointing out that this tenet reflects a dysteleological philosophy. However, at a recent meeting of the NABT in Minneapolis the board has agreed to drop the words "unsupervised" and "impersonal."[96]

NABT executive director Wayne W. Carley said, "We decided that we had construed a meaning we had not intended. [The Statement] was interpreted to mean we were saying there is no God. Absolutely not. We did not mean to imply that. That is beyond the purview of science."[97] NABT member Eugenie Scott added that the revision was "a matter of staying religiously neutral."[98] This is more evidence against Johnson's view that "modern science educators" are *"absolutely insistent* that evolution is an unguided and mindless process."[99]

---

[94] See Eugenie Scott, "NABT Statement on Evolution Evolves," found at the following Internet site <http://www.natcenscied.org/nabt.htm>.

[95] National Association of Biology Teachers, "Statement on Teaching Biology," *American Biology Teacher* 58 (1996): 61.

[96] This revision is certainly a welcome improvement, but leaving the qualifier "unpredictable" may prove to be problematic for some who may construe this natural process as dysteleological.

[97] Reported by Ira Rifkin (Religion News Service): "Teachers Change Evolution Wording," in *The Plain Dealer* (Cleveland), October 16, 1997, 10E.

[98] Ibid.

[99] *Defeating* Darwinism, 15.

Denis O. Lamoureux

Response to Denis O. Lamoureux

Phillip E. Johnson

Denis Lamoureux has shown his passion for this subject by the length of his essay. I don't think readers will have the patience to read through an equally long rejoinder, so I'll just reply to the main points.

The Blind Watchmaker Thesis

Romans 1:20 tells us that, "since the creation of the world, God's invisible qualities—his eternal power and divine nature—have been clearly seen, being understood from what has been made." Some of us find confirmation of this principle in the nature of living organisms, which incorporate many highly complex systems whose nature seems to point to the need for a designer. As Richard Dawkins wrote on the opening page of *The Blind Watchmaker*, "Biology is the study of complicated things that give the appearance of having been designed for a purpose."[1] Darwinian evolutionary biologists insist that this appearance of design is an illusion, and that unguided material processes, particularly the accumulation of random genetic changes through natural selection, have actually produced the wonders of the living world. I customarily refer to this central claim of modern neo-Darwinism as the *blind watchmaker thesis*.

The other major claim of evolution is often called the *common ancestry thesis*. In its most comprehensive form, it states that there is a continuous chain of descent from the first living organism to all the organisms alive today. For reasons I briefly explain in *Defeating Darwinism — By Opening Minds* (pp. 94–95) I doubt that the common ancestry thesis is true, at least at the higher levels (phyla) of the taxonomic hierarchy. However, I do not consider this issue to be of central importance and do not attempt to argue the question for now, because certain crucial work in progress that bears on common ancestry has yet to be published. Common ancestry in itself does not do away with the need for a creator. It is the blind watch-

---

[1] Richard Dawkins, *The Blind Watchmaker* (London: Penguin: New York: W. W. Norton, 1986).

maker mechanism, supposedly capable of creating apparent wonders of design without the need for a designer, that (in Dawkins' words) "makes it possible to be an intellectually fulfilled atheist." The scientific part of Denis Lamoureux's critique defends only the common ancestry thesis, and not the blind watchmaker thesis, and hence he engages the central theme of my work only at the philosophical level.

Is the blind watchmaker thesis true? My writings, and those of my colleagues in the intelligent design movement, assume that the truth or falsity of a scientific thesis should be determined by scientific evidence, rather than decided by philosophical presupposition. We are opposed by persons who endorse methodological naturalism, a doctrine that insists that science must explain biological creation only by natural processes, meaning unintelligent processes. Reference to a creator or designer is relegated to the realm of religion, and ruled out of bounds in science regardless of the evidence. Where methodological naturalism is enforced, unintelligent processes are presumed to be responsible for the origin of life and its present complexity. Evidence is relevant only to determining the precise naturalistic method by which biological organisms arose — including such questions as whether chance-driven or law-driven processes were more important. Any notion that an intelligent designer actually played a role in biological creation is dismissed in the very definition of science.

Unsurprisingly, many methodological naturalists are atheists, agnostics, or deists. Some are theists, however, including active Christians who deny that banishing God from science amounts to banishing him from reality. Christian methodological naturalists usually call themselves theistic evolutionists. A few who take the same position prefer the term evolutionary creationists, presumably because it sounds more acceptable to theological conservatives. Putting aside questions of biblical interpretation, the great problem for both groups is to explain what is 'theistic' or 'creationist' about an evolutionary process that employs only unintelligent causes. God-guided evolution would be genuinely theistic, but the doctrine of methodological naturalism rules out the possibility that God did the guiding in any way that is testable. If it is to achieve the objective of reconciling religion and mainstream science, theistic evolution has to look exactly like naturalistic evolution to the objective observer. The theism is in the mind (or faith) of the believer. For this reason, I have written that theistic evolution can more accurately be described as theistic naturalism.

Is the evolutionary creationism of Denis Lamoureux different from what I have just described as theistic naturalism? It might seem so, because he endorses teleological evolution, and even claims, on the basis of polls showing that many scientists believe in a personal God, that teleological evolution is acceptable within mainstream evolutionary science. On closer examination, however, it appears that the 'teleology' part is entirely subjective and has no more scientific content than the 'theism' in theistic evolution. In that case, it does not violate the rules of methodological naturalism. What exactly did God do (beyond establishing the laws at the beginning of time) and how do we know that he actually did it? Lamoureux meets any attempt to explore that question scientifically with the standard God-of-the-gaps objection, the trademark thought-stopper of the theistic naturalist. Assigning God a detectable role in evolution is a fallacy, according to theistic naturalists, because science will eventually produce a naturalistic explanation for whatever God is supposed to have done. To put the same point another way, all statements about God's work in creation must be unfalsifiable, because otherwise they will surely be falsified.

For example, consider how Lamoureux dismisses the problem of irreducible complexity. A biological system is irreducibly complex when its operation requires the cooperation of numerous parts, none of which performs a useful function unless all are present. Such a system cannot be built up one part at a time—unless some purposeful entity is guiding the process. An unintelligent mechanism like natural selection could not create and preserve a presently useless part because of some long-term goal, but a creator with a goal in mind might do so. Recognition of irreducible complexity thus implies a role for a designer, meaning at a minimum some entity capable of pursuing a distant goal.[2]

Lamoureux rules out the possibility of irreducible complexity without considering the scientific evidence. Why? He says that to consider the need for intelligent causes in biology is merely to place a hypothetical God in the gaps of present scientific knowledge. When those gaps are

---

[2] Lamoureux writes that irreducible complexity could not "develop through a natural process like evolution (whether teleological or dysteleological)." This is a mistake, if by "teleological evolution" he means a gradual process guided by an intelligent agent (like God) who is capable of pursuing a distant goal. Irreducibly complex machines like computers are produced just that way, by engineers who assemble them part by part. It is only unguided (Darwinian) evolution that cannot produce irreducible complexity, because it is incapable of preserving a part which is presently useless but could become useful in connection with other parts when they appear. I think Lamoureux overlooks this because the "teleology" in his evolution is without scientific content.

eventually filled with explanations employing only unintelligent causes, as Lamoureux assumes they inevitably will be, God will be expelled from the gaps and theism will be discredited. To argue this way is to commit the fallacy of begging the question. The adequacy of any naturalistic explanation for irreducible complexity is the very point at issue. Like other theistic naturalists, Lamoureux seems to think he has *a priori* knowledge that naturalistic processes, employing only unintelligent causes, were capable of doing all the work of creating biological systems that are far more complicated than spaceships or computers. Such *a priori* knowledge does not come from experimental science or fossil studies; it comes only from naturalistic philosophy.

Lamoureux's evolutionary creationism, like theistic evolution in general, therefore looks exactly like fully naturalistic evolution to objective observers. The evolution is objective knowledge, which all rational people must accept; the theism or creationism is subjective belief, real only to those who have faith. Lamoureux is a genuine theist, in that he tells us that God is active in the world through charismatic experiences, as manifested in his Pentecostal church. Sincere as it undoubtedly is, this way of confining God to what scientists understand as subjective experience fits comfortably with the naturalistic philosophy that dominates the intellectual world. Evolutionary naturalists are not necessarily hostile to religious belief, provided God stays within the realm of the subjective and never invades the territory of science, where we decide for public purposes what really happened.

It is conceivable that God for some reason did all the creating by apparently naturalistic processes, perhaps the better to test our faith, but surely this is not the only possibility. My writings, and those of colleagues like Michael Behe, argue that design is detectably present in biology, that naturalistic substitutes like the blind watchmaker mechanism are inadequate and contrary to the evidence, and that theists who believe that God is real should not assume that he never played a detectable role in biological creation. We have plenty of evidence to offer, but the evidence does not matter if intelligent causes are ruled out of consideration on *a priori* philosophical or theological grounds. For this reason, I do not think it worthwhile to discuss detailed evidentiary questions with Denis Lamoureux, or with other persons who take the position I call theistic naturalism, whatever they choose to call it. Is the presence of intelligent design in biology a legitimate subject for scientific investigation, to be determined on a level playing field by the weight of evidence? If the answer is "yes," we can go on to assess the evidence—meaning all the evidence, not just a few 'proof texts' selected from a vast mass of contrary evidence. If the answer is "no," as to Denis Lamoureux it seems to be, then to discuss evidence is pointless.

Ad Hominem Arguments

An argument is *ad hominem* when it shifts attention from the merits of an argument to direct attention instead at the qualities or status of the person making the argument. Most of Lamoureux's reasoning is either an appeal to the authority of experts, or an *ad hominem* attack on my own qualifications. (These are the positive and negative sides of the same line of argument.) Whether we should defer to the authority of professional biologists depends largely upon the nature of the issue in controversy. Biologists are best qualified to tell us the details of biology, but they are not qualified to tell us what philosophical assumptions we should adopt. Because the commitment to methodological naturalism is foundational to evolutionary science, such details as the differences between various vertebrate paleontology textbooks, and the precise identification of the candidate mammal ancestor for whales are a sideshow, aimed at diverting the discussion away from the main philosophical questions and into a morass of technical details. Practically none of the evidence supports the standard neo-Darwinian picture of macroevolution, when the philosophical bias is removed. Even if a mesonychid did somehow become a whale, nobody knows how such a spectacular transformation could happen by a series of mutations that increased fitness at every stage, as the orthodox theory requires. If Lamoureux seriously disputes this, let him publish the detailed papers showing the mechanism. His career in science will be a glorious one.

The "trust the experts, not the law professor" argument is *ad hominem* in character, but this is not to say it is without force. If Phillip Johnson is right, a lot of highly qualified people have been wrong, and this includes some prominent Christians. How can this be possible? I have heard that objection from many people, and Denis Lamoureux presses it particularly hard. It is thus absurd for him to label my explanation for the confusion among experts as *ad hominem*. To answer my critics by explaining how the thinking of the experts has gone wrong is not to direct attention away from the merits of the argument, because the existence of confusion in the thinking of the experts is the very point in question.

In all three of my books, and in numerous articles, I discuss why the manifest contradictions between Darwinism and the evidence have for so long been invisible to methodological naturalists of both the agnostic and the theistic variety. Very briefly, both have defined science in such a way that it must ignore or suppress evidence that points to the need for a creator. It is no wonder that agnostics endorse this restriction on thought, because they thereby enlist the unimpeachable authority of scientific investigation in support of their religious position. Theistic naturalists go along with this kind of thinking because they believe that they are saving theism by conforming it to science. This would be laudable if they were conforming theism to the results of unbiased investigation, but (as the all-purpose God-of-the-gaps mantra illustrates) they are actually conforming it to a fundamentally naturalistic way of thinking that ensures that evidence of design in biology will always be explained away, however unconvincingly, as the product of unintelligent causes.

How Theistic Evolutionists might further the Discussion

Theistic evolutionists (including those who call themselves evolutionary creationists) have a credibility problem that stems from their apparent willingness to find support for their compatibilism in the most unlikely places. For example, Denis Lamoureux cites a remark from a letter to claim that Darwin himself remained a theist, ignoring the historical record that shows Darwin to have become a self-proclaimed agnostic by the time of his death.[3] Lamoureux also tries to reassure us that there is no conflict between evolutionary science and theistic religion by citing the willingness of evolutionary naturalists to have conferences with tame theists, meaning those who can be relied upon not to question methodological naturalism. Most egregiously, he misinterprets a tactical shift by the National Association of Biology Teachers as if it were a genuine change in their position. Make no mistake about it; the NABT remains dedicated to presenting evolution as an "unsupervised, impersonal, unpredictable, and natural process" that was not guided by God or programmed to reach a particular goal. The NABT removed the first two terms from their official statement because they were too explicit in revealing the philosophical agenda. The purposeless and unguided nature of evolution is still implied throughout the statement, and the more politically astute evolutionary naturalists have always considered it safer to

---

[3] See Adrian Desmond and James Moore, *Darwin: The Life of a Tormented Evolutionist* (New York: Warner, 1991), 479, 636, 657.

pursue their agenda by persistent insinuation rather than by direct statements (which invite refutation).

Theistic evolutionists may think that evolutionary scientists are more friendly to theism than they seem, but they need to make an effort to convince the rest of us that they are not just looking at the world through rose-coloured glasses. Why not address the scientific establishment frankly on these questions? I would be impressed if the National Academy of Sciences issued a statement criticizing Richard Dawkins and his ilk for misusing science to support atheism, and unequivocally affirming that God-guided evolution is a perfectly respectable interpretation of the available scientific evidence. What the National Academy has actually done is to give its Public Welfare medal to Carl Sagan, who was as notorious a promoter of an atheistic worldview as Dawkins himself.

In 1981, while preparing its legal case against the creation-science movement, the National Academy resolved that "Religion and science are separate and mutually exclusive realms of human thought whose presentation in the same context leads to misunderstanding of both scientific theory and religious belief." I offer Denis Lamoureux and any who agree with him this opportunity: go to the National Academy and other organizations that officially represent scientists and science educators. Tell these organizations that even theistic evolutionists agree that Darwinian evolution is being used by many to advance a naturalistic worldview, and that there is a widespread impression that the promotion of this worldview is not only tolerated but approved by the leading scientific organizations. See if you can get these powerful groups to agree that the promoting of metaphysical naturalism in the name of science violates the National Academy's 1981 resolution, and see if you can get them to support a new statement unambiguously disavowing the mixing of scientific and religious claims by (for example) Dawkins and Sagan. I don't mean to imply that such a statement would put an end to all controversy, because there would still be many scientific questions to address. What would be put to rest is the sense that theistic evolutionists are content with the most nominal reassurances from people whose determination to promote a naturalistic worldview is evident from their conduct.

Conclusion

Finally, I urge theistic evolutionists not to be so frightened of intellectual controversies. If there is to be a renaissance of the evangelical mind, evangelicals need to have something interesting to say that is not just a Christianized version of what the evolutionary naturalists discovered first. The book that Lamoureux was supposed to be reviewing is titled *Defeating Darwinism – By Opening Minds.* (He seems to have left off the last part of the title.) I want to teach students to be critical of scientific claims that are based upon naturalistic assumptions. The message of Lamoureux's review is quite different from that: it is to tell Christian students that they should always trust the experts, and never take an intellectual risk. If that attitude is widespread in Christian higher education, it is no wonder that (to quote Mark Noll) "the scandal of the evangelical mind is that there is not much of an evangelical mind."

The Gaps Are Closing: The Intellectual Evolution of Phillip E. Johnson

Denis O. Lamoureux

I am grateful to Professor Johnson for responding to my paper. It is clear that we both reject what he terms the blind watchmaker thesis of evolution (or what I call dysteleological evolution). In addition, we both agree that there exists design in the universe and that it reflects the mind of a Designer. However, we disagree on how this design, in particular that seen in biology, has arisen. Johnson usually appeals to acts of direct divine intervention from outside the universe during geological history to account for it, while I believe that biological design could result from a process (specifically, a teleological evolution) within the universe as ordained and sustained by God. Despite our disagreements regarding the origin of life, this exchange has been beneficial in that I have defined more sharply certain aspects of my position.

Engaging Johnson on Evolution: Lessons in Escape and Evasion

I have noted two characteristics in the response of antievolutionists like Johnson to the challenge offered by Christians who believe that Jesus created life through an evolutionary process. First, if pressed on specific scientific evidence supporting evolutionary theory, they employ an escape-and-evasion argument, and as a result fail to engage in fruitful dialogue. In order to facilitate our discussion, I specifically raised scientific issues that Johnson deals with in his books (e.g., whale evolution) instead of making an appeal to my scientific specialty of the evolution of teeth and jaws. However, *not once* in his rebuttal is the scientific evidence discussed in any substantive manner. Rather, he makes the characteristic sweeping generalization that the scientific data does not support the theory of evolution. Second, Johnson and antievolutionists fail to engage directly with the Christ-ordained and sustained evolutionary model proposed by their Christian colleagues. Their escape-and-evasion tactic involves *subtly* recasting this teleological view of evolution into dys-teleological categories (a position that Christians who accept evolution are vehemently opposed to) and then directing their argument against it, giving the reader the impression they have met the challenge. Note how Johnson begins his response by affirming that I "engage the central theme of [his] work" because I am opposed to the blind watchmaker thesis of evolution. However, in very next sentence he redirects the discussion away from my challenge by asking, "Is the blind watchmaker thesis true?" and then throughout the rest of his rebuttal he focuses on "unintelligent causes/processes/mechanisms" in an attempt to recast my position on origins within dysteleological categories. Johnson's use of these two characteristic responses from the antievolutionary tradition is my chief criticism of his response to my paper.

The escape-and-evasion routine begins as early as the first two sentences of Johnson's response: "Denis Lamoureux has shown his passion for this subject by the length of his essay. I don't think readers have the patience to read through a rejoinder equally as long, so I'll just reply to the main points." Instead of dealing with the merits of my argument or the evidence that I claim supports evolution, Johnson subtly slights me by suggesting that my response is too long and that emotion is the operative force behind it. Of course, the irony of his complaint is that he later appeals to "all three of [his] books," which total nearly 600 pages. I am sure that if readers have the patience to read all of Johnson's books, a response by him equally as long as mine would not prove very taxing to them.

Johnson's most dramatic escape and evasion also occurs early in his response: "I doubt that the common ancestry thesis is true. . . . However, *I do not consider this issue to be of central importance* and do not attempt to argue the question for now, because certain crucial work in progress that bears on common ancestry has yet to be published" (my italics). This is indeed a remarkable statement! Johnson has gained international attention for rejecting biological evolution (or common ancestry), and now he claims that this theory is not "of central importance." In the book that thrust him to the forefront of the modern antievolution movement, *Darwin on Trial* (1991), seven of the twelve chapters attempt to argue directly against the evidence for common ancestry. Note, too, that in not wanting to deal with the scientific evidence for evolution, Johnson's escape-and-evasion tactic includes an appeal to work "yet to be published." What then is the point in having an evolutionary creationist like me in an exchange of ideas with Johnson if the topic of common ancestry is not central to the discussion?

Johnson further employs his escape-and-evasion routine by discrediting my handling of the scientific evidence. He does "not think it worthwhile to discuss detailed evidentiary questions with Denis Lamoureux" because of my purported philosophical bias . . . and "to discuss evidence is pointless" because I limit the evidence to "just a few 'proof texts' selected from a vast mass of contrary evidence." In other words, instead of dealing with the scientific evidence presented in my paper such as the whale fossil series (*Hapalodectes, Pakicetus, Ambulocetus natans, Rodhocetus, Basilosaurus, Protocetes,* and *Zygorhiza*), Johnson deems these to be "proof texts," thus escaping and evading the data, and then concludes that a "philosophical bias" is the basis of my acceptance of evolution. But it is interesting to note that though he does not want to discuss the scientific evidence for common ancestry, in the next paragraph he unapologetically appeals to it with the characteristic sweeping generalization that "practically none of the evidence supports the standard neo-Darwinian picture of macroevolution." The irony of Johnson's argument is that the proof texts method is used throughout his critique of vertebrate evolution in *Darwin on Trial* (see chapter 6: "The vertebrate sequence"). The evolution of fish to amphibians is dealt with in half a page, amphibians to reptiles in one paragraph, reptiles to birds in one-and-a-half pages, reptiles to mammals in two-and-a-half pages, and apes to humans in four-and-a-half pages. In other words, in under ten pages he tries to deal with the vertebrate fossil record, a documented record that boasts well over 30,000 described genera. In total, Johnson uses 16 fossil vertebrates in an attempt to refute the evolution of this entire subphylum. By anyone's standard, this is clearly a proof text method.

In his conclusion Johnson complains that "the book Lamoureux *was supposed* to be reviewing is titled *Defeating Darwinism – By Opening Minds*" (my italics; he also seems to have left off the last part of the title). As I note at the beginning of my paper, "before assessing [*Defeating Darwinism*] it is necessary to examine and evaluate the foundational principles that Johnson claims to have established in his first two works." Johnson believes that he "had taken on the scientific evidence for Darwinian evolution in *Darwin on Trial*,"[1] and he simply assumes that this is a fact in *Defeating Darwinism*. As a result, I first had to deal directly with Johnson's earlier arguments before proceeding to review his latest book.

[1] *Defeating Darwinism*, 9.

However, the entire second half of my paper is a review of *Defeating Darwinism*, and Johnson has decided not to respond to my criticisms. For that matter, I even apply Johnson's own methodology (i.e., his so-called "baloney detector") to his book in order to fulfil his vision of opening minds. I believe it is evident that he employs straw-man and *ad hominem* arguments (see his treatment of Emilio, scientists, and Christian theologians), and that there is a valid question regarding his trustworthiness as an expert and authority because he steps beyond the limits of his academic expertise and authority (see his assessment of the HIV/AIDS hypothesis). Disappointingly, Johnson chooses the escape-and-evasion route in not responding to my criticisms of his third book.

In sum, Johnson's rebuttal to my paper characterizes the antievolutionary tradition when faced with the challenge of a Jesus-ordained and Jesus-sustained evolution. More specifically, he fails to engage the scientific issues, and he prefers to redirect his criticism to dysteleological evolution. Moreover, Johnson does not think that readers have the patience to read a full response to my criticism, nor does he believe that common ancestry is "of central importance" to our discussion, nor is discussing evidence with me worthwhile, and finally he claims that I never reviewed *Defeating Darwinism*. It is obvious that Johnson is employing the rhetorical tactic of escape and evasion.

The Intellectual Evolution of Johnson on Evolution

The important question that arises when I read Johnson's response to me is: "What exactly is his position on evolution today?" At first he asserts, "I doubt that the common ancestry thesis is true, at least at the higher levels (phyla) of the taxonomic hierarchy." In other words, Johnson is open to evolution *within* a phylum but not *across* phyla. However, when discussing vertebrate evolution he contradicts himself by firmly stating that "*practically none* of the evidence supports the standard neo-Darwinian picture of macroevolution" (my italics). So which is it? If Johnson considers that evolution within a phylum (e.g., chordates) is viable, then to do so means he is persuaded by a vast amount of the "evidence support[ing] the standard neo-Darwinian picture of macroevolution." This suggests that his position on evolution has significantly evolved since 1991 when he finished writing *Darwin on Trial*. For example, in that book's most important chapter Johnson concluded that the fossil record for vertebrates (which are only a *subphylum* of chordates)

Denis O. Lamoureux

weighs "heavily against" the theory of common ancestry.[2] But his openness to evolution within a phylum today argues that he must now consider the evidence for the common ancestry of the vertebrate subphylum feasible; specifically, the evolution of fish to amphibians to reptiles to birds and mammals, including apes to humans. In other words, Johnson's current intellectual evolution has him on a path that betrays the central thesis of *Darwin on Trial*, the book that made him internationally famous.

Johnson's expanding knowledge of whale evolution further points to his intellectual evolution. As I noted in my paper, his recognition of the ancient whale *Ambulocetus natans* in *Defeating Darwinism* is proof of his growing understanding of this fossil series beyond the impression he gave in *Darwin on Trial* that *Basilosaurus* is the only whale transitory form. And now it seems that after reading my paper Johnson is even more open to the "standard Neo-Darwinian picture" for the terrestrial origin of whales, as seen in his speculation: "Even if a mesonychid did somehow become a whale . . . ." That is, it appears that he has (without any apology or recantation) subtly rejected his original claim in *Darwin on Trial* that evolutionists claim that whales descended from rodents.

However, note again the escape-and-evasion technique that Johnson employs immediately prior to this statement regarding mesonychids: "The precise identification of the candidate mammal ancestor for whales [is] a sideshow, aimed at diverting the discussion away from the main philosophical questions and into a morass of technical details." But it must be remembered that this evolution of whales "sideshow" and diversion of the discussion" actually originates in Johnson's *Darwin on Trial*, and that he presses it particularly hard to support his antievolutionism. In the conclusion to his chapter on the vertebrate fossil record in that book, he argues:

> By what Darwinian process did useful hind limbs wither away to vestigial proportions, and at what stage in the transformation from rodent to sea monster did this occur? Did rodent forelimbs transform themselves by gradual adaptive stages into whale flippers? We hear nothing of the difficulties because to Darwinists unsolvable problems are not important.[3]

In responding to this passage in my paper, I point to evidence in the fossil record of the transitional whale series, and I even speculate on a

---

[2] *Darwin on Trial*, 87.
[3] *Darwin on Trial*, 87.

possible mechanism for the transformation of limbs into flippers using developmental biology. It appears that this evidence has influenced Johnson since he is now open to the possibility of the evolution of whales from mesonychids. But more significantly, there is a clear shift in his assessment of the importance of whale evolution to the origins debate. In *Darwin on Trial*, whale evolution was foundational evidence against the theory of common ancestry; now Johnson claims this evidence is simply a "sideshow" and "a diversion in the discussion."

It appears then that Johnson's introduction to the fossil evidence beginning with the terrestrial mesonychid *Hapalodectes*, through the series of ancient whales *Pakicetus, Ambulocetus natans, Rodhocetus, Basilosaurus, Protoceles,* and *Zygorhiza*, has filled a knowledge gap in his understanding of whale evolution, and that his position on evolution is indeed evolving in an evolutionary direction. Such an intellectual dynamic is typical of those holding the God-of-the-gaps position. That is, as more information is gained concerning a purported gap in nature, where God is thought to have once acted, it becomes apparent that the Lord's ordained natural laws and sustenance are sufficient to account for a natural process that transforms one species into another. The closure of the gaps is simply a function of an expanding knowledge base. Johnson's shift in an evolutionary direction testifies to the reality of this intellectual dynamic.

The Reduction of 'Irreducible Complexity': More on the Intellectual Evolution of Johnson on Evolution

The notion of irreducible complexity is foundational to Johnson's view of biological origins. However, as I read his interpretation of this concept in the rebuttal to my paper, the important question that arises is again, "What exactly is his position on evolution today?"

At first, Johnson defines a biological system as *irreducibly complex* if its "operation requires the cooperation of numerous parts, none of which performs a useful function unless all are present." The implication is that such a system could only originate *in toto* because a gradual process such as evolution could not select and keep non-functional parts until the system became completely functional. This is, in fact, the way irreducible complexity was first defined by Michael Behe in *Darwin's Black Box* (1996):

> By irreducibly complex I mean a single system composed of several well-matched, interacting parts that contribute to the basic function,

wherein the removal of any one of the parts causes the system to effectively cease functioning. An irreducibly complex system cannot be produced directly (that is, by continuously improving the initial function, which continues to work by the same mechanism) by slight, successive modifications of a precursor system, because any precursor to an irreducibly complex system that is missing a part is by definition nonfunctional. An irreducibly complex biological system, if there is such a thing,[4] would be a powerful challenge to Darwinian evolution.[5] Since natural selection can only choose systems that are already working, *then if a biological system cannot be produced gradually it would have to arise as an integrated unit, in one fell swoop,* for natural selection to have anything to act on.[6]

Behe's view that irreducibly complex biological systems are assembled in one fell swoop rather than gradually was well-received by antievolutionists, in particular progressive creationists like Johnson and others in the intelligent design movement. That is, Behe's work was happily interpreted to imply that divine intervention at some point in geological time was necessary to account for the assembly of these integrated units since their origin is purportedly not reducible to the laws of a gradual process like evolution.

However, the *reduction* of irreducible complexity becomes evident in Johnson's response to my paper as it accompanies his intellectual evolution toward an evolutionary view. After defining irreducibly complex systems and claiming that they "cannot be built up one part at a time," he quickly qualifies that an exceptional case would be if "some purposeful entity is guiding the process." Becoming more specific, Johnson then adds that these complex structures could be assembled through evolution if it is "a gradual process guided by an intelligent agent (like God) who is capable of pursuing a distant goal" (footnote 1 in the rebuttal).[7]

---

[4] In contrast to Johnson's confidence, it is interesting to note that Behe, the individual who conceived the notion of irreducible complexity, is cautious on whether "there is such a thing" as an irreducibly complex biological system.

[5] Behe defines "Darwinian evolution" in the following manner: "In its full-throated, biological sense . . . *evolution* means a process whereby life arose from non-living matter and subsequently developed entirely by natural means. That is the sense that Darwin gave to the word, and the meaning that it holds in the scientific community." Michael J. Behe, *Darwin's Black Box: The Biochemical Challenge to Evolution* (New York: Free Press, 1996), x–xi.

[6] *Darwin's Black Box*, 39. My italics.

[7] Johnson misquotes me in this footnote, and as a result he misrepresents my position. In this passage I am not referring to my view (simply because I never employ Behe's category of irreducible complexity), but to that of Johnson and his intelligent design col-

Finally, he appears to go even further by suggesting, "Recognition of irreducible complexity thus implies a role for a designer, meaning at a minimum some entity capable of pursuing a distant goal." If this is indeed the case, then Johnson could accept a God who pursues a distant goal by ordaining and sustaining the laws of the universe for life to evolve, including us who bear his precious image.[8] Ironically, this latter position is exactly the one I uphold and term teleological evolution (*telos* is the Greek word meaning the "*end* or *goal* toward which a movement is being directed"[9]). Moreover, this is the view of Johnson's chief critic, Howard Van Till, who argues for the functional integrity of the universe; that is, that God created a world completely equipped for physical structures and life to unfold without acts of divine intervention ("although not proscribed") in the course of time.[10] So again I ask, which is it? Are irreducibly complex biological structures assembled "in one fell swoop," or part by part through a God-guided gradual evolution, or has God programmed and then sustained nature's laws to evolve life as his "distant goal"?

In asserting the viability of the last two options, it is clear that Johnson betrays the original definition of irreducible complexity as it was first proposed by Michael Behe. Moreover, the ease with which Johnson postulates the assembly of irreducibly complex systems in an evolutionary context confirms his intellectual evolution toward the acceptance of common ancestry.

---

leagues. What I wrote is: "However, Johnson and the design theorists introduce a unique twist to the notion of design. For them design carries an aspect of *irreducible complexity*. That is, they assert that certain biological structures are fashioned in such a way that it was not possible for these to develop through a natural process like evolution (whether teleological or dysteleological)."

[8] In principle, Johnson could go even further by limiting God's activity to simply setting up the laws for life to evolve without his constant sustenance, a view commonly known as "deism." However, this is certainly not the position I and other evolutionary creationists uphold. The notion of a God who creates the world and is an absentee landlord is not consistent with the God of the Holy Scriptures.

[9] W. F. Arndt and F. W. Gingrich, *A Greek–English Lexicon of the New Testament and other Early Christian Literature* (Chicago: University of Chicago Press, 1979), s.v. "*telos*", 811.

[10] Howard J. Van Till, "Basil, Augustine, and the Doctrine of Creation's Functional Integrity," *Science and Christian Belief.* (April 1996) 8: 21–38.

Johnson on the Philosophy of Science:
Presuppositions, God-of-the-Gaps, and Intelligent Causes

Despite the fact that Johnson fails to engage in a discussion of the scientific evidence for evolution, the philosophical issues he raises in his response to me are valuable. More specifically, his philosophy of science has three important features that must be examined — the philosophical presuppositions, the God-of-the-gaps view of creation, and the sharp dichotomy between intelligent and unintelligent causes/processes/mechanisms.

Johnson summaries his philosophy of science as "the truth or falsity of a scientific thesis should be determined by scientific evidence, rather than philosophical presupposition. We [i.e., the intelligent design movement] are opposed by persons who endorse 'methodological naturalism,' a doctrine that insists that science must explain biological creation only by natural processes, meaning unintelligent processes." He then concludes that "Lamoureux seems to think he has *a priori* knowledge that naturalistic processes, employing only unintelligent causes, were capable of doing all the work of creating biological systems that are far more complicated than spaceships or computers. Such *a priori* knowledge does not come from experimental science or fossil studies; it comes only from naturalistic philosophy." However, note again the escape-and-evasion tactic characteristic of the antievolutionary tradition that I describe at the beginning of this paper. Johnson *subtly* recasts my teleological view of evolution into dysteleological categories (i.e., an evolution built only on "unintelligent causes/processes" — a position that I am vehemently opposed to) and then directs his argument against it, giving the reader the impression he has met my challenge.[11]

The important question that emerges from this discussion is: On what philosophical presupposition or *a priori* knowledge does Johnson base his philosophy of science? When he asks, "What exactly did God do (beyond establishing the laws at the beginning of time) and how do we know that he actually did it?" is he not making the philosophical presupposition, based on *a priori* knowledge, that God's creative method requires interventionistic acts during the history of the universe? Note that this assumption, too, "does not come from experimental science or fossil studies." Johnson's philosophical presupposition about direct divine activity is logically no different from presupposing that God did not create interventionistically after "establishing the laws at the beginning

---

[11] See page 55.

of time." Every philosophy of science has built-in assumptions, and the challenge then becomes both their *identification* and *justification*. As I point out in my paper, Johnson fails to identify and justify his bias or philosophical presuppositions to his readers.[12]

Johnson's philosophy of science is based ultimately on a certain theological view; in particular, on an interpretive approach to the early chapters of God's Word that I term cosmological concordism.[13] That is, he assumes that the first two chapters of Genesis reveal that life on earth was created by direct acts of divine intervention over geological time. This God-of-the-gaps (a term that for some reason seems to offend Johnson and his colleagues, but in fact should not) position is logically possible. But whether it is true of God's creative method (or the correct interpretation of the Scriptures) is open to question. If the Lord created in this 'progressive' manner, then the 'gaps' in nature should be identifiable and further scientific research should 'widen' them. That is, the statistical improbability of a natural process accounting for the evolution of a complex biological structure or evolution across phyla should increase when probed scientifically over time.

History though has shown that the God-of-the-gaps position has consistently fallen short. Instead of the 'gaps' in nature getting 'wider' with the advance of science, they have closed. Before the discovery of the theory of gravity, many early scientists believed that God or divine beings moved heavenly bodies along their celestial paths. Similarly, only two hundred years ago the best minds in Europe and America interpreted earth history in terms of numerous catastrophic floods due to divine intervention, but this was all before the principles of geology were discovered (in particular, the effects of glaciation).[14] It is primarily because of the *testimony of history* that I am suspicious of Johnson's God-of-the-gaps philosophy of science; and it is *not*, as he wrongly judges, because of philosophical presupposition or *a priori* knowledge.[15] History shows that this approach has consistently failed every time it is used.

---

[12] See section entitled "Trustworthy Experts and Appeals to Authority" in my paper.

[13] See footnote 79 in my paper for the definition of concordism.

[14] For an excellent review of this history see Davis A. Young, *The Biblical Flood: A Case Study of the Church's Response to Extrabiblical Evidence* (Carlisle: Paternoster Press; Grand Rapids: Eerdmans, 1995).

[15] Please remember that I do not because of a philosophic presupposition discount God's activity in his universe; after all, I have revealed my charismatic orientation in my paper. God can enter his creation at any time or in any fashion he chooses. The question is whether he did so interventionistically in the creation of life.

It is important to recognize that the modern God-of-the-gaps theory is being promoted mostly by individuals who are not scientists (specifically, Johnson and the intelligent design movement); and for the few who are scientists, their primary scientific discipline is not evolutionary biology. If gaps do exist in nature where God has directly intervened in the past, then should not researchers who deal with the primary data daily make this important judgment? In addition, this latest God-of-the-gaps approach continues to claim support from the early chapters of the Book of Genesis. However, history has also shown that such concordist interpretations of the Word of God have on every occasion failed.[16] Those promoting this interpretive approach today (i.e., Johnson and the intelligent design movement) are also not the primary researchers in the discipline of interpreting the early chapters of the Word of God. Among modern evangelical Old Testament scholars, there are few who hold such a concordist interpretation. Unfortunately, evangelical concordists do not seem to have learned the lessons of the Galileo affair; tersely stated, "The Bible tells us how to go to heaven, not how the heavens go." Or put in the context of the present discussion on origins, the Holy Scriptures tell us Jesus is the Creator, not how Jesus created.

Johnson's God-of-the-gaps philosophy of science leads him to create a sharp dichotomy between "intelligent causes" and "unintelligent causes/ processes/mechanisms" in the universe. He claims that the former constitute the "role in biological creation" that God plays, and that they are even "detectable" or "testable" by science. In this context, "intelligent causes" are clearly acts of divine intervention in the universe and consist of "whatever God is supposed to have done . . . beyond establishing the laws at the beginning of time." The unfortunate perception that results from this dichotomous view of nature is that "unintelligent causes/processes/mechanisms," including "chance-driven or law-driven processes" appear in a negative light as second- or lower-class operations in creation. But these are in fact ordained and sustained by God to declare his glory in the Creation.

The artificiality of Johnson's dichotomy in nature emerges if it is applied to developmental biology; specifically to human embryology. Where in our development in our mother's womb did "intelligent causes" or "unintelligent causes/processes/mechanisms" operate? Is the actual entrance of genetic material from a sperm cell into an egg cell due

---

[16] For a excellent review of the many failed concordist attempts, see Stanley L. Jaki, *Genesis 1 Through the Ages* (London: Thomas Moore, 1992).

to an "intelligent cause" (i.e., God's direct intervention) or an "unintelligent cause/process/ mechanism?" How about the first division of the fertilized egg? Or the two-cell stage? Or four-cell stage? Or gastrulation? Or neurulation? Or the first heartbeat? Johnson claims that he is opposed to "methodological naturalism, a doctrine that insists that science must explain biological creation only by natural processes, meaning unintelligent processes." However, is he opposed to the medical specialty of obstetrics and gynecology, a science that is practised under the assumption of methodological naturalism that each of us has been created by natural processes without divine interventionistic acts?

In the last century, an important argument used by evangelical scholarship in coming to terms with evolution was based on the analogy between the evolution of species and the embryological development of individual creatures.[17] Instead of a sharp break in nature between "intelligent causes" and "unintelligent causes/processes/mechanisms," these evangelicals had a *unified vision* of the laws ordained and sustained by the Creator in the creation of life both at the level of all species and of each individual. And it is for this reason that nineteenth-century evangelical scholars could accept teleological evolution as proposed by the science of that time. Even one non-evangelical scientist argued, "To my mind it accords better with what we know of the laws impressed on matter by the Creator, that the production and extinction of the past and present inhabitants of the world should have been due to secondary causes like those determining the birth and death of the individual."[18] Of

[17] See James R. Moore, *The Post-Darwinian Controversies: A Study of the Protestant Struggle to Come to Terms with Darwin in Great Britain and America 1870–1900*. (London: Cambridge University Press, 1979); David N. Livingstone, *Darwin's Forgotten Defenders: The Encounter Between Evangelical Theology and Evolutionary Thought* (Grand Rapids: Eerdmans, 1987); Denis O. Lamoureux, *Between "The Origin of Species" and "The Fundamentals": Toward a Historiographical Model of the Evangelical Response to Darwinism in the First Fifty Years*. PhD dissertation. University of St. Michael's College and Wycliffe College at the University of Toronto, 1991.

[18] Charles R. Darwin, *On the Origin of Species*. Facsimile of the first (1859) edition, introduced by Ernst Mayr (Cambridge, Mass.: Harvard University Press, 1964)), 488. I suspect that because I am appealing to Darwin in my argument, Johnson is again going to charge that I "have a credibility problem, which stems from [my] apparent willingness to find support for [my] compatibilism [i.e., between my Christianity and evolution] in the most unlikely places [i.e., appealing to Darwin's *On the Origin of Species*]). However, it must be underlined that it is because of this compatibilism with Darwin's works that evangelical scholarship in the last century came to terms with evolution. Moreover, Johnson writes, "Denis Lamoureux cites a remark from a letter to claim that Darwin himself remained a theist, ignoring the historical record that shows Darwin to have become a self-

course, this scientist is Charles Darwin and the passage is from his classic, *On the Origin of Species* (1859).

Despite our differences, Johnson and I are strong supporters of natural theology. That is, we believe that "The heavens declare the glory of God" (Ps. 19) and that humanity is accountable because of this revelation in nature since it is "clearly seen, being understood from what has been made, so that men are without excuse" (Rom. 1:20). As a result, we both affirm the power of the argument of design, and see it as a valuable tool in the defence of the Christian faith. However, according to Johnson this design in nature seems to be primarily the result of God's direct acts of intervention. This is why he claims that "assigning God a detectable role in evolution is a fallacy, according to theistic naturalists" like me. For Johnson, God's "detectable role" must "violate" the natural order of the universe. But it is necessary to emphasize that his view, though possible, is not logically necessary. In contrast, I maintain that not only do we see design in nature as it is present before us, but design is also expressed in the processes of nature. In other words, I have an argument from design that is twice that of Johnson and other antievolutionists. The *static argument of design* underlines the complexity seen in structures present in front of us (e.g., the 26 trillion synaptic connections in a human brain) and the *dynamic argument of design* points to the complexity of the proc-

---

proclaimed agnostic by the time of his death." (citing Adrian Desmond and James Moore, *Darwin: The Life of a Tormented Evolutionist* (New York: Warner, 1991: 479, 636, 657). Ironically, it is Johnson who ignores the historical record. First, the passage I cite does not come from a letter—it is in Darwin's *Autobiography* in 1876, six years before his death. Besides, the context of my argument regards the use of the word "Darwinism," and I argue "with the recent and rapid growth in the historical studies on the life of Charles Darwin it is abundantly clear that his view of evolution was not dysteleological, though occasionally he considered that possibility." No matter what Darwin's final position was at the time of his death, it is irrelevant compared to how he presented his views to the public throughout his lifetime. Despite this, my point remains (and Johnson even confirms my contention by stating that Darwin was "a self-proclaimed agnostic") the term 'Darwinism' was not used in a dysteleological context by Darwin—and this is all that I was claiming in my paper. It is Johnson's misrepresentation of the term 'Darwinism' as dysteleological evolution in all his books that I object to. Finally, it is interesting that Johnson suggests that I ignore the historical record. Does he appeal to it? No, he appeals to the secondary literature, a biography by Desmond and Moore, who wrote their book within the confines of social constructivist historiography. This is not to say I do not value Desmond and Moore's contribution because it is a fine resource. However, regarding spiritual matters, in particular the possible interpretation of the Holy Spirit's interaction/conviction in Darwin's life, I suggest that one proceed cautiously because Desmond and Moore do not consider this spiritual reality. Therefore, their interpretation of Darwin's religious development must be read cautiously.

esses and mechanisms in nature that allow these structures to emerge (e.g., the evolutionary and embryological operations that account for a human brain with 26 trillion synaptic connections).

Finally, I must qualify my view of methodological naturalism. Obviously, I do not accept the version of this position as Johnson defines it: "a doctrine that insists that science must explain biological creation only by natural processes, meaning unintelligent processes." Note that Johnson's use of the term "unintelligent" introduces a metaphysical element into his definition. I believe that the processes in nature reflect intelligence and are designed by God, but when I make this assertion I step beyond science into the realm of philosophy and theology. Instead, for most scientists methodological naturalism is simply the approach to understanding and explaining the universe by natural processes without any metaphysical references to whether these are "intelligent" or "unintelligent." Moreover, it must be emphasized that I am not slavishly chained to this method in the sense that I categorically dismiss the possibility of God's intervention into the universe. God can certainly do whatever he wants. And it is logically possible that there are gaps in nature attributable to God's direct action. But the question arises, "Does God routinely intervene to 'violate' the laws he established?" Like many who are intimately involved in science (both in research science as an experimental biologist in developmental biology and in clinical science as a practising dentist) I fail to see the gaps in nature that Johnson claims exist. As I noted earlier, history does not support this God-of-the-gaps view. Moreover, as a charismatic Christian and one who has experienced the reality of 'signs and wonders,' it is clear to me that God graciously and miraculously intervenes in our lives. This is why I believe it is necessary to differentiate between 'cosmological interventionism' and 'personal interventionism.'[19] The former deals with the realm that science explores on a daily basis, recording consistent and regular patterns that allows us to speak of the laws of nature. The latter involves God's dealings with us at a personal level through unique events that by definition are irregular and unpredictable, and as such cannot be categorized into law-like generalizations. As a result, I hold a *qualified methodological naturalism* in that this method is a powerful explanatory approach to understanding the origin and operation of nature, but I am not by definition closed to God's direct intervention, in particular into our lives for his glory.

---

[19] See footnote 79 in my first paper for an elaboration of these categories.

In sum, Johnson's philosophy of science is based primarily on the presupposition that when God first created the universe it was incomplete, and that over time it was absolutely necessary for the Creator to intervene directly in the world in order to finish his creation. This philosophical presupposition of the God-of-the-gaps is logically feasible. However, it is not by necessity the Creator's method of creation. Moreover, history has clearly shown that this God-of-the-gaps assumption has consistently fallen short whenever it has been used by Christians. In addition, this presupposition is not used by those intimately involved with science. Finally, design in nature could arise through a God-ordained and sustained process, and it does not necessarily require direct divine intervention for its existence as proposed by Johnson's dichotomous view of nature.

Johnson and *Ad Hominem* Arguments: What Christian Scholars Are Saying

Johnson claims that a significant part of my critique is "an *ad hominem* attack on [his] own qualifications". I am sorry he feels this way, and I apologize if I have offended him. I have no doubt that Johnson is academically qualified with "a specialty in analyzing the logic of arguments and identifying the assumptions that lie behind those arguments."[20] However, I question whether he is sufficiently informed in the biological sciences and evolutionary theory. Before Johnson can apply his special skills to any topic, it is necessary that he have a foundational knowledge of that topic. To ask about whether he has a reasonable grasp of biology is not an *ad hominem* attack, but a reasonable question and concern because it is a fact that Johnson does not even have an undergraduate education in biology. The many scientific errors I (and many others) have noted in his writings (e.g., his rodents to whales thesis), and his escape-and-evasion tactic in not wanting to deal with the scientific evidence I present, indicate he is not well-versed in this discipline. But more telling is Johnson's intellectual evolution in an evolutionary direction since it reflects his becoming better informed about the biological sciences and evolutionary theory. As a result, and now with evidence from his rebuttal, I remain firm in the conclusion of my paper: "Johnson is simply not familiar with the topic [i.e., biological sciences and evolutionary theory]; and thus the application of his analytical skills and their final results

---

[20] *Darwin on Trial*, 13.

must be deemed suspicious at best, if not outrightly unacceptable." This is not an *ad hominem* argument.

Charges of *ad homenium* arguments are serious allegations. I do not think it is fruitful for Johnson and I to enter into an extended exchange over this accusation. Instead, let us examine a recent issue of *Christianity Today* (8 December 1997), the most important popular evangelical journal in North America and one sympathetic to Johnson's views, to see what evangelical scholars are saying about Johnson's treatment of Christian academics. Population biologist David Wilcox of Eastern College notes that if one holds the Christian faith and evolutionary biology "you're accused of lacking integrity" by Johnson.[21] Physicist Howard Van Till of Calvin College asserts that Johnson "makes a number of very harsh accusations about teachers at Christian colleges, that they have swallowed the presuppositions of naturalism, that they are thirsty for approval of non-scientific peers, that they are willing to do and say anything to maintain funding for research."[22] And in a book review of Johnson's *Defeating Darwinism*, Richard Mouw, the president of Fuller Seminary, reports that in "a recent radio interview with James Dobson he [i.e., Johnson] had some harsh words for evangelical scientists who, he said, are so desirous of gaining 'prestige' in the larger academy that they are willing to follow this 'accommodationist' [i.e., acceptance of evolution] route."[23] Mouw wisely admonishes, "Johnson would do well to consider a more charitable interpretation of his Christian colleagues' motives on this matter." I believe that these observations by leading evangelical scholars are consistent with those I have made in my paper regarding Johnson's use of *ad hominem* arguments against Christian brethren.

Questions for Professor Johnson

One of the problems in any exchange of ideas like this one is that individuals may talk past one another. In order to facilitate our discussion, I would appreciate it if Professor Johnson could answer directly the following questions in his next response:

1. In *Darwin on Trial* you claim there is little scientific evidence for the evolution of the vertebrate subphylum. Yet in your first response to

---

[21] Quoted in Tim Stafford, "The Making of a Revolution," *Christianity Today* 41 (8 December 1997): 21.

[22] Quoted in Stafford, "Making of a Revolution," 21.

[23] Richard J. Mouw, "Science with Baloney Detectors," *Christianity Today* 41 (8 December 1997): 50.

me it is clear that you are open to evolution within a phylum. Have you not rejected a foundational argument in the book that thrust you to the front of the antievolutionary movement?

2. Cite the reference from professional evolutionary literature where you found the rodent-to-whale theory that you refer to in *Darwin on Trial*?

3. Reveal how many university-level courses you have successfully completed in biology, and could you be specific in your answer indicating what type of biology these were (e.g., zoology, botany, microbiology, etc.)?

4. In the light of the recent comments made in *Christianity Today* (8 December 1997) by Drs. Wilcox, Van Till, and Mouw regarding your judgment of Christian scholars who accept evolution as God's method of creation, do you still question the integrity and motives of these Christians?

Conclusion: The Gaps Are Closing

In Johnson's conclusion, he suggests that "if there is going to be a renaissance of the evangelical mind, evangelicals need to have something *interesting* to say that is not just a Christianized version of what evolutionary naturalists first discovered" (my italics). I disagree. In order to glorify our Lord and Saviour, the evangelical mind must first pursue that which is true simply because the intellectual titillation of having "something interesting to say" is not a criterion of truth. If Jesus did indeed create life through an evolutionary process, and if science has discovered evidence of this process, then it behooves us evangelicals to integrate these truths into our theology.

Johnson's second conclusion is that "the message of Lamoureux's review is . . . to tell Christian students that they should always trust the experts, and never take an intellectual risk." Ironically, I am telling Christian students exactly the opposite. I am encouraging them not to trust an expert in legal argumentation who is evangelicalism's most important antievolutionist—Dr. Phillip E. Johnson. I hope to teach these students to be critical of the antievolutionary claims in their faith tradition and the assumptions upon which these are based (e.g., the God-of-the-gaps position, the literalist and concordist interpretations of Genesis 1–11, etc.). With antievolutionism solidly entrenched in evangelicalism, these Christian students can be assured they will be taking an intellectual risk in their colleges, seminaries, churches, and even families.

The greatest problem with the intelligent design movement that Johnson leads is that they have yet to offer a definitive theory of origins. This fact is reflected in his response to me by his apparent intellectual evolution toward an evolutionary model. Johnson is not sure whether there is evolution within a phylum, and I believe his uncertainty is telling in that it points to his limited grasp of the scientific evidence. Similarly, he is not certain how the purported irreducible complexity in biological structures is created, suggesting a wide range of mechanisms from fiat creation to God-guided evolution to even an evolution governed by laws programmed by God. I have followed and interacted with the intelligent design movement ever since the 1994 C. S. Lewis Summer Institute on Science and Religion at Cambridge University.[24] Since that time they have been promising a model of origins, but none has come forth for critical examination. Johnson's response to me is typical of the intelligent design movement. He claims that they "have plenty of evidence to offer" and that "crucial work in progress that bears on common ancestry has yet to be published." But the problem remains—a model is promised, but one is never delivered.

To conclude, *the 'gaps' in Johnson's view of nature are in the process of closing and his position on biological origins is evolving in an evolutionary direction.* As a result, his response to me is confusing and at times contradictory. This is due to the fact that philosophically Johnson is still committed to the presupposition of the God-of-the-gaps, yet his expanding knowledge of evolutionary biology is forcing a closure of the purported gaps in nature. That is, on the one hand, Johnson's philosophy of science demands the existence of gaps in nature that can only be filled by intelligent causes; specifically, by the direct intervention of God. On the other hand, the scientific evidence for common ancestry has contributed to his openness to evolution within a phylum and to his reduction of irreducible complexity, resulting in a closing of the gaps. I agree with Johnson that "the truth or falsity of a scientific thesis should be determined by

---

24 At that conference in a lecture by Johnson's colleague Michael Behe (author of *Darwin's Black Box*), I well remember philosopher–theologian William Lane Craig's frustration after his thrice-repeated question, "Well, if evolution isn't true, how then did life arise?" On every occasion, Behe never even attempted to offer Craig an answer. This anecdote typifies the problem with the intelligent design movement—they dismiss evolution, but offer no alternate model.

Acknowledgments: Many thanks to those who have reviewed and commented on this paper: Michael Caldwell, Stewart Carson, Jeremy Lafreniere, Larry Martin, Cory McGlone, Sue Zukiwsky and Ted Zukiwsky.

scientific evidence, rather than decided by philosophical presupposition." And indeed, it is the scientific evidence that Johnson has been ex-
posed to since the time he first wrote *Darwin on Trial* (1991) that is closing
the gaps in his view of biological origins.

Many of us have gone through this process of the 'closing-of-the-
gaps' before Johnson. The evangelical debate over biological origins is
principally an information problem, or rather a *lack* of information problem . . . and devoting ourselves to reading the scroll of God's works is the
most important decision that we evangelicals can make in coming to
terms with evolution, Jesus' method of creation.

Final Response to Denis O. Lamoureux

Phillip E. Johnson

T he important points are adequately covered in my prior response. I'll add only one item: the development of the embryo in the womb is in no way comparable to evolution, because embryonic development is directed by pre-existing genetic information transmitted from the parents. Embryonic development starts with the genetic information already present; evolution has to explain the origin of the genetic information. Similarly, a computer does its work on the basis of a pre-existing program, but an intelligent agent is required to design the program. Failing to observe this distinction is a fallacy I call Berra's Blunder, and I explain the fallacy in *Defeating Darwinism – By Opening Minds* (see pp. 62–64).

I should probably also add for the record that I disavow Denis Lamoureux's characterizations of what I have written. Readers who want to know my views must get them from my own writings.

The two Lamoureux pieces are typical of the criticism I receive from theistic evolutionists, both in substance and in tone. I invite readers to decide for themselves whether that is the kind of intellectual leadership they want the evangelical world to have.

Readers may also be helped by these references:

The profile of me in *Christianity Today* (December 8, 1997) is by Tim Stafford. It may be read on the Web at <http://www.arn.org/johnson/revolution.htm>. The profile in its entirety conveys a very different impression from what one might assume from the selective quotes used by Denis Lamoureux. I take pride in the widely known fact that debates with me, however vigorous, usually end with mutual good feeling and even new friendships.

The review of *Defeating Darwinism – By Opening Minds* by Fuller Seminary President Richard Mouw appeared in the same issue of *Christianity Today*, and may be read on the web at <http://www.christianity.net/ct/7TE/7TE048.html> Lamoureux correctly quotes Mouw as stating in this review that "in a recent radio interview with James Dobson he [Johnson] had some harsh words for evangelical scientists who, he said, are so desirous of gaining 'prestige' in the larger academy that they are willing to follow this 'accommodationist' route. Mouw also wrote that "I should make clear at the outset that I agree with the substance of his [Johnson's] case against the evolutionist perspective as he defines it." He called for a "sustained conversation" about evolution and naturalism among Christian professors. I think that is a very promising idea, and I hope that President Mouw himself will take a leading role in seeing that such a conversation occurs at Christian institutions such as Fuller Theological Seminary.

The following is the transcript of the relevant part of the radio interview (Sept. 15,1997) to which President Mouw referred. JCD is Dr. James C. Dobson; PHIL is myself.

JCD: We spoke a few minutes ago in my office before we came down here, Phil, and I was very gratified to learn that you're doing a lot of speaking, that the doors are open to you on university campuses and in churches, denominations, that people are more and more interested in this perspective. It's gaining credibility. But I want to ask you something that's really delicate, and I hope people understand my motive here.

My wife and I believe in Christian education. Both our kids went to Christian elementary schools and Christian high schools and both graduated from Christian colleges. I believe in Christian higher education and have supported it very strongly. But I heard you say when you came here to speak that you have taken this message—challenging Darwin and challenging classical evolution—onto some of the most prestigious university campuses in the western world, and yet, it has been resisted—your message has been resisted—more on some Christian campuses than on the secular university campus. How do we explain that?

PHIL: Well, in the first place, I want to say it's not universal. There are Christian college professors among my closest associates, and some of them are really on top of this. So that's the good news. The bad news is that in Christian higher education, there is a lot in the faculty culture—it's not coming from the trustees or the presidents; it's in the faculty culture—a lot of resistance to raising the questions that I'm raising.

The basic reason for this, as I see it, is that a generation has grown up which has seen it as their mission to make peace between Christianity and the secular academic organizations. Indeed, even their own constituency—the parents and the trustees and so—want them to educate their young people so they can go on to Stanford and Berkeley and Harvard for graduate school, for example. And they want their professors to have good reputations in the secular academic organizations, you know, of professors. So to get that you have to think like the secular academic people, and so various ways have been found to say that, "Well, Christianity even of an evangelical sort can be reconciled with the way the secular thinking is done." And the biggest area of this is theistic evolution. Because, you see, if you are typed in the leading circles as challenging evolution, then you become a fundamentalist, anti-intellectual, an enemy of science and so. So, you can't have that respect and prestige.

So, one Christian college professor put it in writing. He said, "If we listen to Johnson, the gap between the academy and the sanctuary will grow wider." You see. And so we must shut this out and not listen to it. How I responded to that was by saying that if the secular academy is founded on naturalistic and materialist thinking, then there ought to a very wide gap between it and the sanctuary—meaning the church, meaning the Christian mind. And I believe we have to raise that issue and challenge it.

So, I'm very . . . as you know, I feel very strongly about this, and I'm going to continue to say it. I want to have those people for friends. I don't want to make war on them, because they're often very, very good people. But this issue is too important not to face squarely.

JCD: I want our listeners to know that you go into these university settings, and you take on the most well-known and published evolutionists in their own den and come out whole.

PHIL: Oh, I do. I enjoy it. We have a battle often, but sometimes it's friendly, not always. Some people get very, very bitter and angry about this. But, you see, my job is to open minds, and I want to go into those secular universities and the Christian ones, too, and say, "Think about this. Put these issues on the table. If you are willing to face them honestly, then you don't have to come to the same conclusions I do. That's okay. You're advancing the program if you just look at the issues squarely." It's the ones who want to avoid the issues that I don't get on with. [end of partial transcript]

Intelligent Design:
The Celebration of Gifts Withheld?

Howard J. Van Till

The works by Phillip E. Johnson have generated considerable attention in evangelical Christian circles. Unfortunately, however, I believe that portions of the message conveyed in these books constitute a step backward on the pathway of articulating a more scientifically informed expression of Christian belief regarding the Creation. The evangelical Christian community does indeed have every right to evaluate the enterprise of scientific theorizing regarding the formational history of the Creation, but I strongly disagree with Johnson's attempt to identify the scientific enterprise with a commitment to the atheistic worldview of naturalism. Evangelical Christians also have a responsibility to evaluate their own concepts of God's creative work in the light of what has been learned about the Creation through scientific investigation, but I do not see the *intelligent design* concept, as that terminology is employed by Johnson and his fellow theorists, as a fruitful means toward that good end.

In my brief remarks in the present context I see no need to engage Professor Johnson's works in great detail. Denis Lamoureux has done an admirable job of that, and I judge that Johnson's response has done nothing to discredit the critique offered by Dr. Lamoureux. My remarks here will be of a more general nature on the larger picture in which these differing perspectives must be evaluated.

Against Evolution, but in Favour of What Alternative?

The books by Johnson join many other antievolution works that have been written by persons outside the professional scientific community. A common claim in this literature is that anyone with a working knowledge of the general principles of argumentation from a body of evidence should be able to do a creditable job of evaluating scientific theories in the light of empirical investigation. Some writers even claim to be in a position to do so with greater objectivity than is possible within the professional scientific community. Johnson's broad claim is that he has in fact done this in regard to the concept of biotic evolution, and that he has thereby earned the right to announce with confidence that this major family of scientific theories should now be discarded.

This is a remarkably bold claim. It presumes that a person who is neither professionally trained in evolutionary biology nor a participant in the day-to-day enterprise of empirical research in that diverse and highly technical field nonetheless has the requisite knowledge and evaluative skills to convincingly discredit the judgment of an entire professional community. I suppose that in principle such a state of affairs is not absolutely impossible, but I must say that I believe the odds for it being the case are even poorer than those for winning the ten million dollar grand prize in a magazine sweepstakes.

Having claimed that he has discredited the paradigm of biotic evolution, what does Johnson offer in its place? What specific scenario of either natural phenomena or divine action does Johnson offer as an alternative to the theories he has chosen to discard? In brief, no specific concept; only an arrow pointing in the direction of what he chooses to call *intelligent design*.

What Does It Mean to Be 'Intelligently Designed'?

One of Johnson's fundamental claims is that he and his associates can point to clear empirical evidence that specific categories of life forms or biotic subsystems have been 'intelligently designed.' Given the diversity of meanings that such a term is taken to connote by persons both within and outside of the scientific and religious communities, it would seem to me imperative that the term *intelligent design* be candidly and publicly defined. In the absence of a clear definition, wasteful confusion rules the day. For that reason I have often asked the proponents of intelligent design to state candidly and publicly exactly what they mean by the term that labels their movement. I am still waiting for that request to be honoured.

Without being provided with a candidly stated definition, we are likely to infer a meaning from the term itself. In contemporary usage, the term 'designed' most commonly draws our attention to the ideas of planning and purpose. Stated more formally, we would ordinarily say that to be designed is to be thoughtfully conceptualized for the accomplishment of some identifiable purpose. As an action, therefore, 'design' is the intentional and purposeful action of a mind.

As I see it, however, the effecting of a design, or the actualization of what was first thoughtfully conceptualized, constitutes a distinctly different sort of action. Hence, in a modern manufacturing system there is one team of persons in charge of designing (planning, conceptualizing) something and another team of persons in charge of assembling (fabricating, actualizing) what has been designed. In modern usage, 'design' and 'assembly' are distinctly different concepts. One of my greatest frustrations with the intelligent design literature is its persistent failure to give recognition to this fundamental distinction, thereby making it nearly impossible to evaluate each of the two concepts, along with relevant empirical evidence, on its own merits.

An earlier usage of the term 'designed' (as it was most commonly employed in the eighteenth century by English clergyman William Paley, for instance) was based on the artisan metaphor. One person, the artisan, did both the conceptualization and the fabrication of some artifact. Paley's watchmaker did both the conceptualization and the construction of the watch. Paley's Designer (like his watchmaker) had both a mind (to conceptualize or intend) and the divine equivalent of 'hands' (to assemble, fabricate, or actualize).

My reading of numerous books and essays written by Johnson and other proponents of intelligent design leads me to the conclusion that the operative (but not candidly or publicly stated) meaning of 'intelligently designed' is essentially the same as the one conveyed by Paley's watchmaker metaphor. Johnson rejects what he calls the 'blind watchmaker thesis', not because he finds any shortcomings in the artisan metaphor, with its conflation of two distinctly differing kinds of action, but because he judges that the artisan must be able to see what specific acts of form-imposing or assembling are required in order to bring about (by means that go beyond the capabilities of the materials or systems on hand) the actualization of some novel creature or biotic subsystem.

To the majority of those who are proposing it, then, to be 'intelligently designed' means to be *both* (1) thoughtfully conceptualized for the accomplishment of a purpose, *and* (2) assembled (at least for the first time) by the direct, form-imposing action of some extranatural agent. In other words, the intelligent design perspective is a form of episodic creationism, a concept of the formational history of the Creation that includes, as an essential element, the idea that certain creatures or biotic subsystems could have been assembled for the first time only by episodes of direct divine intervention in which novel biotic forms were imposed on some extant materials or biotic systems.

All Christians believe that the Creation was thoughtfully conceptualized for the accomplishment of God's comprehensive purposes. But not all Christians, including Denis Lamoureux and myself, believe that the formational history of the Creation to which God gave being, and continues to give being, was punctuated by episodes of form-imposing divine intervention. Having a wholehearted commitment to the historic Christian doctrine of creation does not in any way commit a Christian to favouring the mode of extranatural assembly that the concept of intelligent design entails.

It is obvious that the intelligent design proposition being zealously promoted by Johnson and most of his associates (there is some disagreement among them on this matter) flows from an interventionist concept of divine creative work and from an empirical evidentialist apologetic strategy for engaging the preachers of naturalism. Curiously, however, one finds no explicit or substantive theological treatment of these foundational presuppositions or strategies in the intelligent design literature.

To What Kind of Creation Did God Give Being?

I see little promise of any substantive progress in the larger discussion as long as the present 'creation versus evolution' format is allowed to set the tone of the exchange. In effect, this simplistic either/or format appears to give the preachers of naturalism a distinct advantage over all forms of episodic creationism, including intelligent design. In order to see how I come to this rather unconventional position, let me here outline briefly my general approach to these issues.

Rather than constructing my position as a reaction to naturalism, I choose to begin by affirming my commitment to the historic Christian doctrine of creation. I do indeed see the entire universe—every atom, every physical structure, every living organism—as a Creation that has being only as an expression of God's effective will. God is the universe's Creator in the fundamental sense of being the One who has given the universe its being. The being of the universe is radically dependent on the effective will of its Creator.

Essential to the being of the Creation is not only a set of properties that characterize its multifarious substances, structures, and organisms, but also a vast array of creaturely capabilities for action and interaction, including capabilities for self-organization and transformation. Elemen-

tary particles called quarks, for instance, possess the capabilities to interact in such a way as to form nucleons (protons and neutrons). Nucleons,
in turn, have the capacities to interact and organize, by such processes as
thermonuclear fusion, into progressively larger atomic nuclei. Nuclei and
electrons have the dynamic capability to interact and organize into atoms. On the macroscopic scale, vast collections of atoms interact to form
the inanimate structures of interest to astronomy — galaxies, stars, and
planets. On the microscopic scale, atoms interact chemically to form
molecules; molecules interact to form more complex molecules. Some
molecular ensembles might well possess the capabilities to organize into
the fundamental units that constitute living cells and organisms. Organisms and environments interact and organize into ecosystems.

All of these organizational and transformational capabilities together
comprise what I have come to call the *formational economy* of the universe.
The universe's formational economy is truly astounding. The natural sciences have only just begun to uncover the remarkable and varied capabilities that comprise it. Each day more of these formational capabilities
are discovered.

But there are two fundamentally differing concepts regarding the
character and robustness of the universe's formational economy. One
concept constitutes the fundamental presupposition of all forms of episodic creationism. The other functions as the basis for the consideration
of non-episodic scenarios for the formational history of the Creation (or
of the universe, if you prefer).

Using the terminology I have just introduced, the question now at issue is this: Is the formational economy of the Creation sufficiently robust
to bring about the actualization of all of the diverse physical structures
and life forms that have appeared in the course of time?

Fifteen centuries ago Augustine proposed a "Yes" answer to this
question. He envisioned God bringing the whole Creation into being in
one comprehensive act and gifting its fundamental substances with all of
the capabilities (he called them 'seed principles') for actualizing the full
array of creaturely forms in the course of time. In his particular scenario
these forms became actualized, not in the sequentially connected manner
of evolution, but independently and contemporarily.

Like Augustine long ago, modern natural science also proceeds on
the assumption that the universe is equipped with a robust formational
economy. This strategy has proved remarkably fruitful, especially in the
arena of the physical sciences like geology, astronomy, and cosmology in
their concern for reconstructing the formational history of the elements,

of space-time, of planet Earth, and of stars and galaxies. Numerous un-solved puzzles remain, of course, but the pattern of progress in account-ing for the formation of a broad spectrum of physical structures is abun-dantly clear. The 'robust formational economy principle' continues to demonstrate its fruitfulness as the basis for the scientific reconstruction of formational histories of physical structures from minuscule atomic nuclei to gigantic spiral galaxies.

What about the formational history of life forms? This presents the biological sciences with questions of considerably greater difficulty than those encountered in the physical sciences, but the same strategy cer-tainly appears to hold great promise. Contrary to Augustine's vision of a side-by-side actualization of the whole array of life forms, however, the biological sciences are now convinced by the evidence that the actualiza-tion of new life forms has occurred in a sequential manner marked by genealogical continuity. As is the case for the physical sciences, biology is a long way from a complete understanding of all of the details of biotic evolution, but nearly every member of the professional scientific com-munity is convinced that the empirical evidence very strongly favours an evolutionary scenario. Some critics have levelled the claim that, because there are still so many gaps in our knowledge of the relevant processes and their particular outcomes, evolution is a "theory is crisis." I think it would be far more fair and accurate to describe it as a "theory in its ado-lescence." It is now in a stage of rapid, ebullient, and sometimes erratic development, but its basic character and potential are clearly visible.

Like other forms of episodic creationism, the intelligent design movement presumes that at least some of today's knowledge gaps (things we do not now know) regarding evolutionary development must be taken as evidence for corresponding functional gaps (caused by missing capabilities) in the formational economy of the Creation. Some proponents of the anti-evolution perspective make the even bolder claim that they can now point to convincing empirical evidence for the absence of certain creaturely capabilities for self-organization or transformation, thereby opening up gaps in the Creation's formational economy that could have been bridged only by form-imposing episodes of assembly by intelligent design.

But why would there be gaps in the formational economy of the Creation? In the present context there can be only one answer to this question. If there are gaps in the Creation's formational economy, it must be by God's choice. God must have chosen to withhold certain forma-tional gifts from it. If Johnson, the intelligent design theorists, and other episodic creationists are correct, then God must have chosen to equip the Creation with an incomplete formational economy—still a remarkable set of gifts, but lacking certain key formational capabilities that would later make necessary a succession of episodes of form-imposing interventions in the course of time. If, on the other hand, God chose to gift the Creation with a robust and gapless formational economy, then either Augustine's side-by-side actualization or the sequential actualization envisioned by contemporary natural science could represent the manifestation of God's intentions for the formational history of the Creation.

For Christians the question is not, Creation or evolution? Rather, the real question is, To what kind of Creation did God give being? One with gaps in its formational economy? Or, one with a robust and gapless formational economy?

The Optimally Gifted Creation Perspective

For a number of reasons I strongly favour the second alternative—the vision of a Creation gifted by God with a robust and gapless formational economy. I call this the 'optimally gifted Creation perspective.' More specifically, since it incorporates the scientific concept of evolutionary development, it could also be called the 'evolutionary creation perspective.' Among the more important factors leading me to a preference for this perspective are the following:

1. Because it is a perspective on the Creation, we are called first of all to recognize that the universe has being only as the outcome and manifestation of the effective will of its Creator-God.

2. Because the universe is a Creation, every property and capability of its creatures (from minuscule quarks to massive galaxies, and from elementary particles to complex life forms) must be recognized as gifts of being that have been given to them by the Creator. From this historic 'creationist' perspective, no natural process, that is, no exercise of creaturely capabilities, may be declared entirely void of 'intelligence' or purpose. For a Christian to do so would, it seems to me, constitute an insult to the Creator who gives being to each and every one of those 'natural' capabilities.

3. The broad concept of a Creation gifted from the outset with a robust and gapless formational economy comports with the heritage of early

Christian thought, such as that represented in Augustine's reading of the first three chapters of Genesis.[1] This vision of an optimally gifted Creation effecting the Creator's intentions for its formational history at all times stands in bold contrast to all episodic creationist scenarios in which "purposeless, unintelligent, naturalistic processes" are only occasionally punctuated by episodes of special creation or intelligent design. The intelligent design concept as it has been promoted by Johnson, with its intense emphasis on episodes of form-imposing intervention and its frequent association of material processes with naturalistic causes, could perhaps be more accurately called a theory of *punctuated naturalism*.

4. In the context of this vision of the giftedness of the Creation, we have reason to welcome every creaturely capability discovered by the natural sciences (the systematic investigation of creaturely phenomena) and to celebrate these gifts as manifestations of the Creator's unfathomable creativity and unlimited generosity. Who but the Creator could give being to a universe so richly gifted?

5. In the expectation that the Creation has been optimally gifted with a robust and gapless formational economy, we will not be tempted to search for empirical evidence of gifts withheld, as if God's creative work would best be known not by the gifts of being that are present but by those that are absent. Furthermore, it would be highly misleading to label the scientific methodology that follows from this high view of the Creation's giftedness with the derogatory epithet 'methodological naturalism,' thereby suggesting that naturalism deserves ownership of contemporary scientific methodology. The worldview of naturalism has no way to account for the existence of anything, certainly not of a Creation gifted with a robust and gapless formational economy.

6. Holding to an optimally gifted Creation perspective would enable the Christian community to avoid the inverted scoring system of the creation–evolution debate, in which every scientific discovery that gives support to the robust formational economy principle is credited to the worldview of naturalism, and the credibility of Christian theism is

---

[1] Howard Van Till, "Basil, Augustine, and the Doctrine of Creation's Functional Integrity," *Science and Christian Belief,* 8 no. 1 (April 1996): 21–38.

made to appear as if it were dependent on demonstrating the existence of gaps in the formational economy of the universe. I can see why the preachers of naturalism love this scoring system, but why would a Christian allow such a travesty to continue? For a Christian to concede ownership of the robust formational economy principle to the preachers of naturalism would be a blunder of colossal proportions. Furthermore, repeatedly to label those Christians who embrace the concept of an optimally gifted Creation with the oxymoronic term 'theistic naturalism' is to demonstrate a reckless disrespect for persons wholly committed to the historic Christian doctrine of creation.

7. Finally, this perspective offers Christians a means by which the Christian theological enterprise may benefit from the informed judgment of the professional scientific community, which includes a large number of Christians, regarding the nature of the Creation and the character of its formational history. This would, I believe, provide theologians with the occasion for articulating numerous theological questions to which the specific character of the Creation is relevant. In the context of intelligent design and other forms of episodic creationism, many of these important questions are, unfortunately, being neglected.[2]

---

[2] For persons who would like to read more material from this perspective, the following is a list of other relevant publications by Howard Van Till:

1. "Can the Creationist Tradition be Recovered?" an essay review of the book *Creation and the History of Science,* written by Christopher Kaiser, *Perspectives on Science and Christian Faith,* 44, no. 3 (September 1992): 78–85.
2. "God and Evolution: An *Exchange,*" with Phillip E. Johnson, *First Things,* no. 34 (June/July 1993): 32–41.
3. "Special Creationism in *Designer* Clothing: A Response to the Creation Hypothesis," an essay review published in Perspectives *on Science and Christian Faith,* 47, no. 2 (June 1995): 123–31.
4. "No Place for a Small God," published as a chapter in *How Large is God?* ed. John Marks Templeton (Philadelphia: Templeton Foundation Press, 1997).
5. "The Creation: Intelligently Designed or Optimally Equipped?" to be published in the October 1998 issue of *Theology Today,* pp. 344-64.

Teleological Evolution:
The Difference it Doesn't Make

Stephen C. Meyer

Denis Lamoureux supports what he calls a teleological view of evolution. He equates the power of God with natural laws and processes, and claims that these laws and processes are goal-directed and creative. Lamoureux favours the view that "God organized the Big Bang, so that the deck was stacked"[1] to favour the inevitable evolution of life by natural law. He says further that he supports natural theology, but unlike other theists who see God revealing himself in the design of specific living systems or finely tuned conditions, he sees evidence of God's handiwork in the natural laws and processes that he alleges can account for the origin of biological systems. Further, Lamoureux objects to Phillip Johnson's critique of Darwinism, because he thinks by focusing his critique on the dominant materialistic view of evolutionary theory, Johnson gives short shrift to the version of evolutionary theory that Lamoureux favours. Though Johnson does not commit himself to a specific scenario involving divine action at a specific point in natural history, Lamoureux criticizes him for what he takes to be Johnson's tacit commitment to some form of Genesis-based special creation. According to Lamoureux, to invoke a specific instance of intelligent design or divine action within nature after the initial creation of the universe would imply a violation of natural law and commit the God-of-the-gaps fallacy. Indeed, Lamoureux claims that design theorists generally (including Johnson) base their arguments for design on God-of-the-gaps–style arguments from ignorance—a claim he amplifies by asserting that design theorists have failed to produce a rigorous theory of intelligent design.

---

[1] Quoted in Joe Woodard, "The End of Evolution," *Alberta Report* December 1996, 33. See footnote 7 below.

I disagree with Lamoureux's position for two main reasons. First, I think that he mischaracterizes design theory and the arguments of design advocates, including those of Phillip Johnson. Secondly, I think Lamoureux's notion of teleological evolution lacks either explanatory power or theoretical specificity or both.

In the first place, contemporary design theory does not constitute an argument from ignorance or a God-of-the Gaps fallacy as Lamoureux claims. Design theorists infer design not merely because natural processes cannot explain the origin of such things as biological systems but because these systems manifest the distinctive hallmarks of intelligently designed systems — that is, they possess features that in any other realm of experience would trigger the recognition of an intelligent cause. For example, Michael Behe[2] has inferred design not only because the gradualistic mechanism of natural selection cannot produce irreducibly complex systems, but also because in our experience irreducible complexity is a feature of systems known to have been intelligently designed. Indeed, whenever we see systems that possess irreducible complexity (i.e., systems with many functionally integrated but necessary parts) and we know the causal story about how they originated, intelligent design invariably played a role. Thus, Behe infers intelligent design as the best explanation for the origin of irreducibly complexity in such things as cellular molecular motors, based upon what we *know*, not what we do not know, about the causal powers of nature and intelligent agents, respectively.

Similarly, Phillip Johnson (following Charles Thaxton and Walter Bradley[3] and me[4]) has argued that the specified complexity or information content of DNA and proteins implies a prior intelligent cause, again because 'specified complexity' and 'high information content'[5] constitute

---

[2] Michael Behe, *Darwin's Black Box: The Biochemical Challenge to Evolution* (New York: Free Press, 1996).

[3] Charles B. Thaxton and Walter L. Bradley, "Information and the Origin of Life" in *The Creation Hypothesis: Scientific Evidence for an Intelligent Designer*, ed. J. P. Moreland (Downers Grove, Ill.: InterVarsity Press, 1994), 173–210.

[4] Stephen C. Meyer, "The Explanatory Power of Design: DNA and the Origin of Information," in *Mere Creation: Science, Faith and Intelligent Design*. ed. William A. Dembski (Downers Grove, Ill.: InterVarsity Press, 1998), 114–47; "The Origin of Life and the Death of Materialism," *The Intercollegiate Review* 31 no. 2 (1996): 24–43; "DNA by Design: An Inference to the Best Explanation for the Origin of Biological Information," *Rhetoric and Public Affairs* (Lansing, Michigan: Michigan State University Press) 1, no. 4 (1999): 519–555.

[5] The term 'information content' is used variously to denote both specified complexity and unspecified complexity. Yet a sequence of symbols that is merely complex but not specified functionally (such as 'wnsgdtej38ejdfmcksdnenmd') would not necessarily indi-

a distinctive hallmark (or signature) of intelligence. Indeed, in all cases where we know the causal origin of high information content or specified complexity, experience has shown that intelligent design played a causal role. Thus, when we encounter such information in the bio-macromolecules necessary to life, we may infer—based upon our *knowledge* of established cause–effect relationships—that an intelligent cause operated in the past to produce the information necessary to the origin of life. Design theorists infer a past intelligent cause based upon present knowledge of cause-and-effect relationships. Inferring design thus employs the standard uniformitarian method of reasoning used in all historical sciences. These inferences do not constitute arguments from ignorance any more than any other well-grounded inferences in geology, archaeology, or paleontology—where provisional knowledge of cause–effect relationships derived from present experience guides our inferences about the causal past.

Indeed, as William Dembski has recently demonstrated, we often infer the causal activity of intelligent agents as the best explanation for events and phenomena. Moreover, we do so rationally, according to objectifiable, if often tacit, information and complexity theoretic criteria. His groundbreaking new book *The Design Inference,* published by Cambridge University Press, gives a formal theoretical account of the criteria by which specialists in many fields reliably detect intelligent causes. Dembski shows that whenever events are both highly improbable and specified,[6] we infer intelligent design (not chance, law, or some combination of

---

cate the activity of a designing intelligence. Thus, it might be argued that design arguments based on the presence of information commit a fallacy of equivocation by inferring design from a type of 'information' (i.e., unspecified information) that could result from random natural processes. Ambiguities in the definition of information and information content do leave open this possibility. One can foreclose this possibility, however, by defining information content as equivalent to the joint properties of complexity and specification. Though the term is not used this way in classical information theory, it has been – used this way by biologists from the beginning of the molecular biological revolution. As Sarkar points out, since the mid-1950s Francis Crick and others have equated 'information' not only with complexity, but also with what they called "specificity"—where they understood specificity to mean "necessary to function." This response will also use the term "information content' to mean specified information, or specified complexity, not just complexity. See Sahotra Sarkar, "Biological Information: A Skeptical Look at Some Central Dogmas of Molecular Biology," in Sahotra Sarkar, ed., *The Philosophy and History of Molecular Biology: New Perspectives* (Dordrecht: Kluwer Academic Publishers, 1996), 191.

[6] In the most general sense, a specification is a pattern or description of an event that is conditionally independent of the event that it describes. Thus, an event is specified if it conforms to a conditionally independent pattern or description. In both a linguistic and

the two) as the best causal explanation for the event or artifact in question. Thus he shows that design inferences are based upon the *presence* of particular features implying an intelligent cause, not (solely) upon the absence of evidence for the efficacy of natural causes. We would not say, for example, that an archaeologist had committed a 'scribe-of-the-gaps' fallacy simply because he inferred that an intelligent agent had produced an ancient hieroglyphic inscription. Instead, we recognize that the archaeologist has made an inference based upon the *presence* of a feature (namely, high information content or small probability specification, to use Dembski's terminology) that implies an intelligent cause. We would not say that he had based his inference (solely) upon the *absence* of evidence for a suitably efficacious natural cause.

Because Lamoureux does not seem to appreciate this distinction (at least in the biological case), he mistakenly attributes Johnson's advocacy of design to a misguided biblical hermeneutic. Johnson, he says, believes in special creationism because he is wedded to a flawed interpretation of Genesis chapters 1–11. But this is incorrect. Neither Johnson, nor any of the other design theorists mentioned above, base their case for design on the book of Genesis. For design theorists, design is not a deduction from religious authority, but an inference from biological and/or physical evidence — indeed, it is an inference underwritten by the very kind of formal theory that Lamoureux mistakenly says the intelligent design movement lacks.[7]

Whether or not design constitutes the best explanation for the origin of biological data may be debatable. But one thing is certain: teleological evolution, insofar as it relies on the laws of nature to create, cannot account for the origin of biological information. Indeed, from an empirical point of view, the laws of nature do not have the creative powers that Lamoureux's position requires. Because Lamoureux and other teleological evolutionists want to limit divine action to the initial creation of the

biological context, a specification is a match or convergence between an event and an independent functional requirement. For a formal account of specification, see William A. Dembski, *The Design Inference* (Cambridge: Cambridge University Press, 1998), 1–66, 136–174.

[7] For other recent technical books advancing the intelligent design position, see Paul Nelson, *On Common Descent*, University of Chicago Evolutionary Monograph Series (Chicago: University of Chicago Press, 1999); William A. Dembski, edt., *Mere Creation: Science, Faith and Intelligent Design.* (Downers Grove, Ill.: InterVarsity Press, 1998); Michael Behe, *Darwin's Black Box.*

universe[8] and its natural laws (and to the maintenance of those laws thereafter), they must rely exclusively on natural laws and processes to explain the origin of biological form and complexity. This includes not only the origin of novel forms from existing forms, but also the origin of the first life itself. Unfortunately, however, the laws of nature lack the power to create the information rich-structures that characterize biological organisms. Natural laws may well be maintained and have been created by God, as all theists (including Johnson) believe, but the physical and chemical regularities that scientists describe as laws do not (by definition) produce the information-rich configurations of matter that the origin of life requires. God may have created natural law, but He does not use natural laws to create specified biological information.

To see why consider the problem of the origin of the first life from simple chemistry. Teleological evolutionists, committed as they are to the proposition that the laws of nature as originally designed by God are sufficient to produce life, must argue for some form of self-organizational origin-of-life scenario. Typically, these scenarios suggest that the forces of chemical necessity (as described by physical and chemical law) make the origin of life and the genetic information that it requires (in DNA, RNA and proteins, for example) inevitable.[9] To a materialist such self-organization suggests the self-existence and creative self-sufficiency of natural law. To a teleological evolutionist such as Lamoureux, it shows the goal-directed nature of natural laws as originally designed by God. And, of course, both these views are logically possible approaches to explaining the origin of life. Both are contradicted, however, by empirical evidence and information-theoretic considerations.

As the physical chemist Michael Polanyi showed in 1967,[10] the laws of physics and chemistry leave open (or indeterminate) a vast ensemble of possible configurations of matter, only very few of which could have any role in a functioning biological organism. Specifically, he noted that

---

[8] Note, for example, Lamoureux's criticism of progressive creationism in a recent issue of the *Alberta Report*: "But progressive creationism, the theory that God had to intervene at different steps along the way—from matter to life—suggests that he couldn't get it right the first time. But if he's God, why couldn't He have organized the Big Bang, so that the deck was stacked entirely in favour of life? So that the intelligent design was already built into matter? This would be evolutionary creationism, more in line with what the fossil record suggests" (quoted in Woodard, "The End of Evolution.")

[9] See, for example, Christian De Duve, "The Beginnings of Life on Earth," *American Scientist* 83 (1995): 437.

[10] Michael Polanyi, "Life's Irreducible Structure", *Science* 160 (1968): 1308–1312, esp. 1309.

the chemical laws governing the assembly of the chemical subunits in the DNA molecule allow a vast array of possible arrangements of nucleotide bases, the chemical 'letters' in the genetic code. In other words, the chemical properties of the constituent parts of DNA (and the laws governing their arrangement) do not determine the specific sequencing of the bases in the genetic molecule. Yet, the specific sequencing of the nucleotide bases in DNA constitutes precisely the feature of the DNA molecule — namely, its information content — that origin of life biologists most want and need to explain.[11]

Since the elucidation of the DNA structure by Watson and Crick in 1953,[12] it has become clear that the coding regions of DNA possess the same property of 'sequence specificity' or 'specified complexity' or 'information content' that written codes or languages do.[13] Just as the specific arrangement, not the chemical properties, of the letters in this article account for the communication function that it performs, so too does the specific sequencing of the nucleotide bases in DNA account for the function that DNA performs within the cell. In particular, the specifically sequenced nucleotide bases on the DNA direct the process of protein synthesis in the cell. The origin of this specific sequencing in DNA represents a mystery for all current chemical evolutionary models of the origin of life and defies explanation by reference to the self-organizational chemical laws that teleological evolutionists must necessarily favour. To say otherwise is like saying that the law-like forces of chemical attraction governing ink on this page are responsible for the sequential arrangement of the letters that give this article meaning.

To see why in more detail consider the following. The accompanying diagram shows the chemical structure of DNA. It shows that this structure depends upon several chemical bonds, each of which are governed by laws of chemical attraction. There are chemical bonds, for example, between the sugar and the phosphate molecules that form the two twisting backbones of the DNA molecule. There are bonds fixing individual (nucleotide) bases to the sugar-phosphate backbones on each side of the molecule. There are also hydrogen bonds stretching horizontally across the molecule between nucleotide bases making so-called comple-

---

[11] Berndt-Olaf Kuppers, *Information and the Origin of Life* (Cambridge, Mass.: MIT Press, 1990), 170–172; also Charles Thaxton, Walter Bradley and Roger Olsen, *The Mysteries of Life's Origin* (New York: Philosophical Library, 1984), 24–38.

[12] James Watson and Francis Crick, "A Structure for Deoxyribose Nucleic Acid", *Nature* 171 (1953): 737–738.

[13] Charles Thaxton and Walter Bradley, "Information and the Origin of Life", 173–210.

mentary pairs. These bonds, which hold two complementary copies of the DNA message text together, make replication of the genetic instructions possible. Most importantly, however, notice that there are *no* chemical bonds between the bases along the vertical axis in the centre of the helix. Yet it is precisely along this axis of the molecule that the genetic instructions in DNA are encoded.[14]

Further, just as magnetic letters can be combined and recombined in any way to form various sequences on a metal surface, so too can each of the four bases, A, T, G, and C, attach to any site on the DNA backbone with equal facility, making all sequences equally probable (or improbable) given the laws of physics and chemistry. Indeed, there are no differential affinities between any of the four bases and the binding sites along the sugar-phosphate backbone. The same type of 'n-glycosidic' bond occurs between the base and the backbone regardless of which base attaches. All four bases are acceptable, none is preferred. As Berndt-Olaf Kuppers has noted, 'a present day understanding of the properties of nucleic acids indicates that all the combinatorially possible nucleotide patterns of a DNA are, from a chemical point of view, equivalent.'[15] Thus, 'self-organizing' laws or properties cannot explain the sequential ordering of the nucleotide bases in DNA because: (1) there are *no* chemical bonds between bases along the message-bearing axis of the molecule, and (2) there are no *differential forces of attraction* between the backbone and the various bases that could account for variations in sequencing.

For those who want to explain the origin of life as the result of self-organizing properties or natural laws intrinsic to the material constituents of living systems, these rather elementary facts of molecular biology have devastating implications. The most logical place to look for self-organizing chemical laws and properties to explain the origin of genetic information is in the constituent parts of the molecules carrying that information. But biochemistry and molecular biology make clear that law-like forces of attraction between the constituents in DNA (as well as RNA and protein)[16] do not explain the sequence specificity of these large information-bearing biomolecules.

---

[14] Bruce Alberts et al., *Molecular Biology of the Cell* (New York: Garland, 1983), 105.

[15] Bernd-Olaf Kuppers, "On the Prior Probability of the Existence of Life," in *The Probabilistic Revolution*, ed. Lorenz Kruger et al. (Cambridge, Mass.: MIT Press, 1987), 364.

[16] R. A. Kok, J. A. Taylor, and W. L. Bradley, "A Statistical Examination of Self-Ordering of Amino Acids in Proteins," *Origins of Life and Evolution of the Biosphere*, 18 (1988), 135–142.

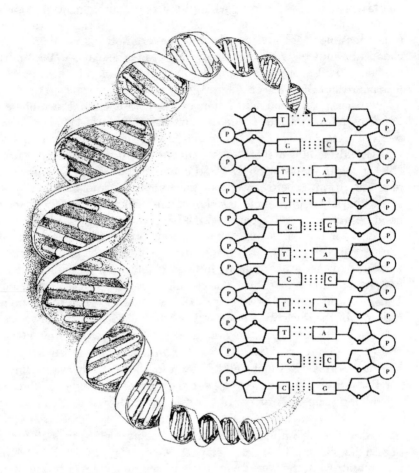

Figure 1: The bonding relationships between the chemical constituents of the DNA molecule. Sugars (designated by the pentagons) and phosphates (designated by the circled Ps) are linked chemically. Nucleotide bases (the A's, T's, G's and C's) are bonded to the sugar phosphate backbones. Nucleotide bases are linked by hydrogen bonds (designated by dotted double or triple lines) *across* the double helix. But *no* chemical bonds exist between the nucleotide bases along the message-bearing spine of the helix. Thus, chemical properties (and the laws that describe them) do not determine the informational sequencing of the DNA molecule. (Diagram used by permission of Fred Hereen and Daystar Productions.)

These facts also raise a very difficult question for teleological evolutionists such as Denis Lamoureux and Howard van Till. Both Lamoureux and van Till insist that God's direct, discrete or special creative activity played no role after the initial moment of creation at the Big Bang. Both imply, therefore, that the laws of nature acting on elementary particles were sufficient to organize matter into the complex forms we see today. Yet if the chemical subunits of DNA lack the self-organizational properties, or latent creative potential, necessary to produce the informational sequencing of DNA, it is difficult to see how the far less complex and biologically specific elementary particles (present just after the Big Bang) possessed the intrinsic properties and potential necessary to arrange themselves (by natural law) into fully functioning organisms. Where are the self-organizational laws and properties that can explain the assembly of the subunits in sequence-specific DNA molecules? Where are they for functioning proteins? Or for signal transduction circuits? Or molecular motors? Where are they for the many specific and highly improbable arrangements of matter that characterize the structures of living organisms? These chemical laws and properties clearly do not exist in the chemistry of the DNA molecule. And if they don't exist there, it seems implausible in the extreme to assert that such creative law-governed properties existed in far simpler constituent parts of atoms or elementary particles.

Lamoureux and other teleological evolutionists might, of course, object that any such negative argument constitutes a God-of-the-gaps fallacy or an argument from ignorance. 'Never say never,' they say. Yet this objection fails here. There are strong in principle information-theoretic objections to any attempt to attribute information-rich structures or sequences to the laws of nature. Scientific laws describe (by definition) highly regular phenomena or structures, ones that possess what information theorists refer to as redundant order. To say that the processes that natural laws describe can generate complex informational sequences is essentially a contradiction in terms. The patterns that laws describe are necessarily highly regular, repetitive, and periodic. But the arrangements of matter in an information-rich text, including the genetic instructions on DNA, possess a high degree of complexity or aperiodicity, not redundant order.

To illustrate the difference compare the sequence ABABABABABAB to the sequence "One small step for man, one giant leap for mankind." The first sequence is repetitive and ordered, but not complex or informative. The second sequence is not ordered, in the sense of being repetitious, but it is complex and also informative. The second sequence is complex because its characters do not follow a rigidly repeating, law-bound pattern. (It is also informative because, unlike a merely complex sequence such as 'sretfdhu&*jsa&90te', the particular arrangement of characters is highly exact or specified[17] so as to perform a (communication) function. In any case, informative sequences have the qualitative feature of complexity (aperiodicity), and thus are qualitatively distinguishable from systems characterized by periodic order that natural laws generate.

Both teleological evolutionists and self-organizational theorists claim that we must await the discovery of new natural laws to explain the origin of biological information. Manfred Eigen has argued, "our task is to find an algorithm, a natural law, that leads to the origin of information."[18] But this claim betrays a categorical confusion. According to classical information theory, the amount of information present in a sequence is inversely proportional to the probability of the sequence occurring. Yet laws necessarily describe highly deterministic or predictable relationships between conditions and events. Laws describe patterns in which the probability of each successive event (given the previous event and the action of the law) approaches unity. Yet information content mounts as *im*probabilities multiply. Information is conveyed whenever one event among an ensemble of possibilities (as opposed to a single necessity) is specified. The greater the number of possibilities, the greater is the improbability of any one being specified, and the more information is transmitted when a particular possibility is specified. If someone tells you that 'it is raining,' he will have conveyed some meaningful information to you since it does not rain (or have to rain) every day. If, however, he also tells you that 'today the raindrops are falling down, rather than up,' he will not have told you anything informative since, presumably, you already know that rain always falls down (by natural law). As Dretske has explained:

> As p(si) [the probability of a condition or state of affairs] approaches 1 the amount of information associated with the occurrence of si goes to 0. In the limiting case when the probability of a condition or state of affairs is unity [p(si)=1], no information is associated with, or generated

---

[17] See footnote 5 above for a definition of specification; see also Dembski, *The Design Inference*, 1–66, 136–174.

[18] Manfred Eigen, *Steps Toward Life* (Oxford: Oxford University Press, 1992), 12.

by, the occurrence of si. This is merely another way to say that no information is generated by the occurrence of events for which there are no possible alternatives.[19]

Natural laws describe situations in which specific outcomes follow specific conditions with high probability. Yet information is maximized when just the opposite situation obtains, namely, when antecedent conditions allow many possible and improbable outcomes. Thus, to the extent that a sequence of symbols or events results from a predictable law-bound process, to that extent the information content of the sequence is limited or effaced (by redundancy). Natural laws do not generate complex informational sequences. Thus, they cannot be invoked to explain the origin of information, whether biological or otherwise.

Of course, explaining the origin of genetic information by reference to laws or properties of physical–chemical necessity does not exhaust the logical possibilities. One can also invoke contingency or chance, either of the directed or undirected variety. For example, many chemical evolutionary theorists have invoked chance alone or chance in conjunction with pre-biological natural selection in an attempt to explain the origin of information. Neo-Darwinists invoke random mutations in DNA to explain the origin of novel biological information in pre-existing organisms. Some theistic evolutionists, such as Gordon Mills, have also suggested mutations as mechanisms of evolutionary change, but have suggested that the improbability of producing functional variations via such mutations is so great as to suggest that these apparently random events must have been directed by a guiding intelligence.[20]

Unfortunately, none of these approaches represent live options for the committed teleological evolutionist such as Lamoureux. For by invoking contingency or chance, whether singly or in conjunction with natural law, the teleological evolutionist runs the risk of qualifying his position out of existence. If, for example, the teleological evolutionist seeks to avoid the information-theoretic difficulties discussed above by invoking *undirected* chance to explain the origin of genetic information, his position becomes indistinguishable from standard materialistic versions of evolutionary theory (either biological or chemical) that Johnson and many others have criticized on empirical, methodological, and theological

---

[19] Fred Dretske, *Knowledge and the Flow of Information* (Cambridge, Mass.: MIT Press, 1981), 12.

[20] Gordon Mills, "Similarities and Differences in Mitochondrial Genomes: Theistic Interpretation," *Perspectives on Science and Christian Faith* 50, no. 4 (December 1998): 286–292.

grounds. (In any case, it should be noted that neo-Darwinism has failed every bit as much as chemical evolutionary theory to provide a mechanism that can explain the origin of specified genetic information — whether the information required to build novel genes, cell types, organs, molecular machines, developmental programs, or body plans that have arisen during the history of life on earth). If, on the other hand, the teleological evolutionist invokes directed contingency (i.e., the active and intelligent guidance of genetic variation during the course of biological history), then he violates his own injunction against employing divine action or intelligent design as a cause during the history of life. At that point the teleological evolutionist will have violated Van Till's doctrine of the "functional integrity of creation" every bit as much as any design theorist. He will also have committed the very kind of argument that Lamoureux explictly repudiates as a God-of-the-gaps fallacy.

Thus, barring an empirically unsupportable and theoretically incoherent commitment to the view that the laws of nature can create novel-specified information, it is difficult to see what empirical content Lamoureux's teleological evolution has or how it differs in substance from standard neo-Darwinism with its denial of any evidence of actual, as opposed to merely apparent, design. To cite C. S. Peirce's maxim "for a difference to be a difference, it must it make a difference." One must ask: does the 'teleological' in the phrase 'teleological evolution' make any scientific difference? No one doubts that Lamoureux and Van Till believe in God as designer. But for that designer to play some role in a scientific theory — such as Lamoureux's teleological evolution — the designer must also play some discernible role in the history of life. The Creator must do something. Yet Lamoureux is unwilling to specify any such role (beyond the causally necessary but insufficient one of creating or maintaining natural law) lest he violate his own self-imposed prohibition against invoking divine action. But if God's activity remains forever superfluous or undetectable (except through the eyes of faith), then it also becomes scientifically irrelevant. Thus, Johnson has, unfortunately, been right to give Lamoureux's version of evolutionary theory short shrift.

Comments on
Denis Lamoureux's Essays

Michael J. Behe

Arguing with a friend is always unpleasant, but especially so when it arises more from misunderstanding than from substantive differences. Denis Lamoureux's essay is full of intensity and emotion, but ultimately it rests on confusion about what other people are saying.

Intelligent Design

The critical point that sceptics about Darwinian evolution such as Phillip Johnson and myself are trying to focus on is the question of intelligent design. Suppose you are taking your morning walk around the block and see that several of your neighbours' yards contain yellow flowers. In one yard the yellow flowers — tulips — grow neatly around the mailbox. In the next yard the yellow flowers are dandelions, scattered hither and thither. You quickly conclude that the first neighbour had purposely planted the tulips, most likely to beautify his home. You decide with equal alacrity that the second neighbour did not arrange the dandelions on purpose; they just happened to grow that way.

How do we recognize intelligent design? We ourselves are intelligent beings, acting to accomplish our purposes, and we are adept at discerning signs of intelligent activity. But how do we do that? How do we distinguish a cultivated flower bed from a patch of weeds? a chance meeting from a conspiracy? radio noise in space from a message sent by aliens? a random universe from an arranged one?

Up until now, detecting design had been a rather intuitive affair. As has been said of pornography, we couldn't define design, but we knew it when we saw it. Spurred on by developments in a number of branches of science, however, the question of design is beginning to receive serious attention. Cambridge University Press has recently published *The Design Inference* by the mathematician/philosopher William Dembski, the first book to treat design as a rigorous academic subject.[1] Very briefly, Dembski says we reach a conclusion of design when we see "specified small probability," a highly improbable arrangement that fits a significant pattern. The tulip bed, for example, is quite unlikely to happen by itself, and acts to grace the neighbour's property. A series of radio waves from space, coding for the first hundred prime numbers, is quite unlikely and would be taken as evidence of intelligent design (the US government's SETI [Search for Extraterrestrial Intelligence] program used such a criterion).

Developing rigorous criteria for design is an interesting intellectual exercise, but what if you get more than you expected? What if you find that not only was the tulip bed designed, but the tulip, too? These are questions that are facing modern science as we dig deeper into the structure of the universe and life. A number of physicists have written on 'anthropic coincidences', features of the universe that seem suspiciously fine-tuned to allow life. It has been shown that if fundamental features such as the charge on the electron or the mass of the proton differed from their measured values by a tiny fraction, the universe could not have produced life. For example, Stephen Hawking writes:

> The remarkable fact is that the values of [the charge on the electron and the ratio of the masses of the proton and the electron] seem to have been very finely adjusted to make possible the development of life.[2]

And the astronomer Fred Hoyle (an atheist, by the way), after observing the felicitous placement of the nuclear resonance levels of beryllium and carbon, declared:

> A common sense interpretation of the facts suggests that a superintellect has monkeyed with physics, as well as with chemistry and biology, and that there are no blind forces worth speaking about in nature. The numbers one calculates from the facts seem to me so overwhelming as to put this conclusion almost beyond question.[3]

Notice that Hoyle is making a design conclusion based on William Dembski's criterion of specified small probability. The likelihood of the resonance levels being where they are is quite low, and is specified to allow the occurrence of life.

[1] William A. Dembski, *The Design Inference* (Cambridge: Cambridge University Press, 1998).

[2] Stephen W. Hawking, *A Brief History of Time* (New York: Bantam Books, 1988), 125.

[3] Fred Hoyle "The Universe: Past and Present Reflections," *Engineering and Science*, November 1981.

Michael J. Behe

Compelling design arguments can also be made about the problem of the origin of life, as well as about the occurrence of systems of what I have called 'irreducible complexity' in the cell. The details of these arguments can be found elsewhere and we need not bother with them here. The point I want to emphasize is that the apprehension of design can be intellectually rigorous, completely empirical, based solely on physical data, and is well within the bounds of science.

A Different Basis for Design

Now, with a foundation in place, let's turn to Denis's argument. He writes, "I believe that the processes in nature reflect intelligence and are designed by God." So Denis believes in design. But why does he believe in it? Does he have the same kinds of reasons that you had when you walked by your neighbor's flower bed? Apparently not, for he continues, "but when I make this assertion [of design] I step beyond science into the realm of philosophy and theology." We generally don't think we're stepping into the realm of philosophy and theology when we recognize a flower bed as designed. We wouldn't even think that we were being philosophical if we decided that radio waves from space were carrying a message sent by aliens.

I think the hitch is that Denis is basing his conclusion of design on a criterion other than specified small probability; he is basing it on authority. It turns out that a second way to conclude that something has been designed is for someone to tell you of the design. So if, for example, a good friend tells you that your co-workers at the office are planning a surprise birthday party for you, you may believe that collusion (design) is going on, even though you see no physical evidence of it. In other words, you take your friend's word for it.

Denis and many other Christians believe in design because they believe in a Designer. They have been told the Good News by someone, or have read the Bible for themselves, and believe that the universe is designed based on God's Word. That, of course, is an excellent reason. However, it is not the same as coming to a conclusion of design based on physical evidence. Furthermore, if you believed your friend's story about the surprise birthday party, you might expect that, if you investigated hard enough, you'd find evidence of the plan being implemented. You might see co-workers coming out of party-goods stores or making calls to the bakery for a cake.

Denis is in the position of someone who says he believes his friend's story, but also claims that a thorough investigation reveals not a trace of physical evidence for the putative party. That is why he retreats to the ethereal realms of "philosophy and theology." Now, it is possible that one might not find evidence for a party if the co-workers were trying very hard to cover their tracks. (After all, they want to surprise you.) And God, of course, could act effortlessly so that we could not discern his work through physical evidence, if he so desired. However, we have no reason to think God wanted to cover his tracks, and the Scriptures say in a number of places that the design of creation is supposed to be quite evident. Thus we reach the peculiar position where the design of creation, based on physical evidence, is less evident to Denis than the design of a neighbour's flower bed, where the design of creation is more apparent to atheist Fred Hoyle than to Christian Denis Lamoureux.

Misplaced Objections

I'm no mind reader, but I think Denis has put himself in this odd position because of concern over other, unrelated issues. First, he believes that the idea of intelligent design based on physical evidence necessarily involves direct intervention by God in the natural world.[4] That is simply not the case. An all-powerful God could set up Creation to unfold as he sees fit. As an analogy, let's think of the pool-sharp Minnesota Fats. If Fats wanted to get six billiard balls in six pockets, he wouldn't have to place them there by hand; he could shoot the cue in such a way that the laws of nature, after the initial hit, produce the desired result. But either way, by hand or by cue shot, we could detect the intelligent intent from the physical result.

Second, Denis appears to think that intelligent design in biology necessarily rules out common descent. That is not true either. In fact, in my book *Darwin's Black Box: The Biochemical Challenge to Evolution* I explicitly

---

[4] For example, apparently referring to me, he writes that "To account for the existence of these irreducibly complex structures, intervention from outside the normal operation of the universe is claimed to have occurred during the history of life." I have never said that. In fact, I am always quite careful to point out that, although most people will conclude that the designer is God, the fact of design itself doesn't demand that conclusion (see chapter 11 of my book, *Darwin's Black Box*). As an example, Francis Crick has famously proposed that space aliens might have started life on earth. Whether the designer of life was God or space aliens is not easily settled by scientific arguments.

say that I think common descent remains a reasonable idea.[5] However, I disagree strongly that Darwinian processes can account for biological complexity. To illustrate the difference, consider the development of a fertilized egg. For a developing human, a single cell gives rise to trillions of modified descendant cells, which form the adult organism. That does not happen by Darwinian natural selection, there is an in-built, pre-existing program to produce the baby. The point is that God could have included explicit information in nature to allow the purposeful unfolding of life by non-Darwinian means. (I should add that although God could have done things in this fashion, perhaps he did not, and the idea of common ancestry may require a second look as science progresses.)

Finally, Denis is worried that science will eventually explain by random processes what seems now to be designed, so that "God [will appear] to be forced further and further into the dark recesses of our ignorance." Denis seems to think the best course to avoid this is to surrender in advance: simply assume that random processes did everything and then we won't be upset in the future. He sees this as inevitable since "History . . . has shown that the God-of-the-gaps position has consistently fallen short. Instead of the 'gaps' in nature getting 'wider' with the advance of science, they have closed."[6]

I find his position difficult to understand. How, I wonder, does preemptive surrender change what people will conclude from the apparent lack of physical evidence for God's activity in the universe? It seems to me that logically consistent people will more quickly reach the same point under Denis's scheme as they would if, as he expects, intelligent design theorists are laid low. Furthermore, I think Denis's history of linearly advancing materialistic explanations is simplistic. Consider a few counterexamples. In the nineteenth century many scientists thought life could spontaneously arise from sea mud. But that idea was discarded when the complexity of the cell was discovered. In the early 1950s Stanley Miller thought he had a handle on how life may have originated, but now most scientists will frankly admit we have no explanation for the beginning of life. Scientists used to think the universe was eternal, which fit nicely with materialism. Now it has a beginning, with obvious attendant theological overtones. And the anthropic coincidences that Hoyle and others have discovered only add to materialism's woes. My point is

---

[5] Michael J. Behe, *Darwin's Black Box: The Biochemical Challenge to Evolution* (New York: Free Press, 1996), 5.

[6] See page 58 of this book.

that history does not tell the tale that Denis thinks. The fact that he relies on such a dubious historical argument tends to highlight his lack of scientific arguments for Darwinian evolution.

Appreciation

I appreciate Denis's taking the time to raise these points and provide me the opportunity to show why I think his objections are based mostly on a misunderstanding of what intelligent design theory claims. Perhaps now he can redirect his intensity to helping us elucidate the intelligent design of life.

Design and Purpose within an Evolving Creation

Keith B. Miller

In his response to Lamoureux, Johnson implies that it is the blind watchmaker mechanism that is the object of his criticism, not common ancestry. But evangelical Christians who are evolutionary creationists (or continuous creationists) also fully reject such a view of evolution. In fact, it is the false equation of metaphysical naturalism with evolutionary theory, which Johnson uncritically accepts, that is the philosophical foundation for the blind watchmaker. By accepting that evolution and metaphysical naturalism are inseparable, Johnson allows the atheists to define the terms and set the agenda of the debate. Evolution becomes an alternative to Christian theism, and the debate is reduced to a choice between undirected, purposeless change and a personal, creatively active God. Our purpose should be to destroy false equalities and false dichotomies, not perpetuate them.

I want to emphatically agree with Lamoureux's assessment that for many, if not most, practising scientists 'evolution' does not mean what Johnson claims it does. Johnson defines evolution as "unguided, purposeless change" that God has nothing to do with. Athough this understanding of evolutionary theory is widely promoted by some atheistic scientists, its conflation of metaphysical naturalism with evolution is widely rejected as philosophically, theologically, and historically false, and is recognized as damaging to the discipline of science. Eugenie Scott, the director of the National Center for Science Education (NCSE), a strong secular advocacy group for the teaching of evolution, states "If science is limited to explaining the natural world through natural processes, we are then constrained from making pronouncements about the supernatural world. We can neither say there is, nor say that there is not, a God or any other omnipotent power. . . . Statements about whether God exists, or interferes in the world, are just plain outside of our job description as scientists, regardless of our personal theistic or nontheistic views."[1]

Clearly Johnson's critique goes far beyond the notion of the blind watchmaker to the viability of evolutionary theory itself. Many of his strongest words are directed not at atheists but at fellow believers. He states that "God-guided evolution would be genuinely theistic," but then proceeds to harshly criticize those who hold just such a view. He appears to do this because of his strongly held position that God's creative activity must be scientifically testable. This view runs the danger of reducing a biblical doctrine to a testable scientific hypothesis. Why should our doctrine of creation be made subject to scientific verification? He also seems to equate any meaningful divine action with breaks in chains of cause-and-effect processes. I see no scriptural justification for this. God's creative activity is clearly identified in Scripture as including natural processes. Thus Christians should not fear causal explanations. Complete scientific descriptions of events or processes should pose no threat to Christian theism, unless a person's theology or apologetic has been inseparably welded to an interventionist view of God's action.[2]

Seeking scientific evidence for divine action in the failure of present scientific description actually has the effect of diminishing the perception of God's action in the physical universe. The very designation of only certain events or structures as 'intelligently caused' relegates all others to the status of 'unguided natural processes.' The argument from design should not be reduced to searches for gaps in scientific description. God is personally active in all natural processes, and all of creation is purposefully designed by God. All creation declares the glory of God. The evidence for God's presence in creation, for the existence of a creator God, is precisely those everyday natural events experienced by us all. The trees, the animals, the seas, flood and storm, the very rocks all proclaim God's reality to anyone who desires to see. It is for this reason that Paul declares that all humanity is without excuse. That is natural revelation. To reduce it to gaps in our scientific explanation does, I believe,

---

[1] From a web article by Scott in which she argues that evolution does not imply metaphysical naturalism, and that theologically loaded terms have no place in scientific description. Eugenie C. Scott (February 20, 1998) "Are Terms Like 'Impersonal' and 'Unsupervised' Scientific? A Personal Commentary on Methodological Materialism," at <http://www.NatCenSciEd.org/publs/ philresp.htm>

[2] See Alvin Plantinga, "Methodological Naturalism?" *Perspectives on Science and Christian Faith* 49 (1997): 143–154. While Plantinga argues against methodological naturalism as a necessary part of science, he is quite emphatic in rejecting any kind of interventionist or God-of-the-gaps understanding of God's action.

great disservice to the witness of God in creation. As I have said elsewhere, "If a person cannot see God in a sunset or a thunderstorm, he or she will not see him in a strand of DNA or a mitotic spindle."[3] The argument from design is that God is praised and revealed through *all that he has made*. My objection to the arguments of the proponents of intelligent design is not that they posit design, but that they restrict its meaning to only certain structures or processes and make it subject to scientific verification.

Johnson consistently characterizes evangelical Christians who accept evolutionary theory as allowing for only unintelligent causes and as having an understanding of God that is basically deistic. Such a characterization could not be farther from my view, in which *all* natural processes are the personal, purposeful act of a creator God. The following are some of my confessional statements regarding God's creative action. God is the creator of all things. All things were created through the Word who is Christ the crucified (John 1:1-3, Col. 1:15-20). God is both transcendent over creation, and immanent in creation. God's creative power is continually at work, even to the present day (Ps. 104:29-30). God is as active in natural events as in miraculous ones. God is intimately and actively involved in what we perceive as natural or law-governed processes (Amos 4:6ff.). God is in control of random or chance events (1 Kings 22:17-38; Acts 1:21-26). God is revealed in the present creation (Job 38-41). God is active in the world, providing for the needs of its creatures (Job, Ps. 104, Matt. 6:25-30). If one accepts the above theological statements, then it seems to me that a completely seamless evolutionary history of life would be entirely acceptable theologically. In other words, such a scientific description would not violate one's understanding of the nature and character of God. This is, and has been, the position of many evangelical theologians and scientists since the time of Darwin.

Since the publication of *The Origin of Species*, evolution has been viewed by many theologically orthodox Christians as a positive contribution to understanding God's creative and redemptive work. This group included even some of the central figures in the early fundamentalist movement.[4] For many, important theological truths concerning the nature of humanity, the goodness of creation, God's providence, and the

[3] Keith B. Miller, "God's Action in Nature," *Perspectives on Science and Christian Faith* 50 (1998): 75.

[4] David N. Livingstone, *Darwin's Forgotten Defenders: the Encounter between Evangelical Theology and Evolutionary Thought* (Grand Rapids: Eerdmans, 1987); James R. Moore, *The Post-Darwinian Controversies* (Cambridge: Cambridge University Press, 1979).

meaning of the Cross and suffering find renewed significance and ampli-
fication when applied to an evolutionary view of God's creative work. [5]

Underlying much of what Johnson and others have written is a sense
that evolution provides a convenient excuse for nontheists to ignore
Christian claims—to be "intellectually fulfilled atheists." Certainly, a
universe in which God revealed himself in a scientifically demonstrable
way would seem to have a great apologetic advantage. But is this what
the doctrine of natural revelation means? As my comments above indicate, I am convinced the answer is no. We should not decide questions of
truth by judging their apologetic effectiveness. God is who he is and acts
as he chooses to act, regardless of whether we find it convenient or not.
There is a great desire among Christians to enlist the authority and prestige of science in support of Christian theism, to have an apologetic based
on irrefutable scientific evidence. But what is the value of founding the
apologetic for our faith upon a constantly changing and limited body of
human knowledge? It is as if we have bought into the culture of our time
in elevating scientific knowledge to the final arbiter of all truth. If a truth
claim cannot be made subject to scientific test, then it is not deemed
worthy of our attention. We must reject this restricted view of truth that
is foreign to the Scriptures.

Johnson objects to the methodological naturalism of science. Methodological naturalism is simply a recognition that scientific research proceeds by the search for chains of cause and effect and confines itself to
the investigation of natural entities and forces. Science does not 'assume
away' a creator—it is simply silent on the existence or action of God. Science restricts itself to proximate causes, and the confirmation or denial of
ultimate causes is beyond its capacity. Methodological naturalism places
boundaries around what science can and cannot say, or what explanations or descriptions can be accepted as part of the scientific enterprise.
Science is self-limiting, and that is its strength and power as a methodology.[6] Science pursues truth within very narrow limits. Our most profound questions about the nature of reality, while they may arise from

---

[5] For examples of positive theological responses to evolution, see John Polkinghorne,
Science and Providence: God's Interaction with the World (Boston: Shambhala Publications,
1989), 114; George L. Murphy, "The Paradox of Mediated Creation *Ex Nihilo*," *Perspectives
on Science and Christian Faith* 39 (1987): 221–226; George L. Murphy, "A Theological Argument for Evolution," *Perspectives on Science and Christian Faith* 38 (1986): 19–26; Keith B.
Miller, "Theological Implications of an Evolving Creation," *Perspectives on Science and
Christian Faith* 45 (1993): 150–160.

[6] Robert C. O'Connor, "Science on Trial: Exploring the Rationality of Methodological
Naturalism," *Perspectives on Science and Christian Faith*, 49 (1997): 15–30.

within science, are theological or philosophical in nature and their answers lie beyond the reach of science. I believe that a great disservice to science has been done by nontheists who have ignored its limitations and transformed it into a naturalistic worldview.

Johnson and other advocates of intelligent design argue that methodological naturalism arbitrarily restricts the search for truth. It does nothing of the sort. If God acted in creation to bring about structure A in a way that broke causal chains, then science would simply conclude that "There is presently no known series of cause-and-effect processes that can adequately account for structure A, and research will continue to search for such processes." That conclusion is perfectly accurate, although tentative. But scientific conclusions are always tentative, since human knowledge will always be limited. Any statement beyond that requires the application of a particular religious worldview. Science cannot conclude "God did it." However, if God acted through a seamless series of cause-and-effect processes to bring about structure A, then the continuing search for such processes stimulated by the tentativeness and methodological naturalism of science may uncover those processes. Using an intelligent design approach, the inference of intelligent design would be made, and any motivation for further research would end. Thus intelligent design runs the risk of drawing false conclusions and prematurely terminating the search for cause-and-effect descriptions when none are yet known. Furthermore, how would a gap in the causal chain be discovered unless continuing effort was expended in searching for possible natural causes? Thus even the verification of gaps requires research conducted using the assumptions of methodological naturalism.

Several proponents of intelligent design define 'design' as 'specified small probability.' Aside from the question of whether probabilities can be reliably assigned to unique historical events, the calculation of probabilities is a perfectly permissible activity within methodological naturalism. There need be no philosophical redefinition of science in order to pursue such goals. But this definition of design does nothing to address the concerns outlined above. If highly improbable events are designed, then the implication is that highly probable ones (such as the sun rising and setting) are not. But Scripture is quite clear in stating that it is just such events that God has ordained for a purpose. Furthermore, a person who sees design in small probabilities should find no objection to the evolutionary creationist positions held by those within the evangelical Christian community. From the perspective of evolutionary creationism, God's creative activity is exercised through natural processes such that complete cause-and-effect descriptions are possible. In the case of the history of life, the expectation is that all life is connected by common descent. Such a theistic understanding implies nothing about probability. Since God is actively guiding the process, there is no reason why he could not bring about events of small probability. An argument based on the calculation of probabilities is in effect a restatement of the anthropic principle that arose within a science directed by methodological naturalism. Thus, if design is identified merely as 'small probability' as advocated by Michael Behe and other prominent intelligent design proponents, both Johnson's critiques of methodological naturalism, and his equation of evolution with philosophical naturalism are baseless. Johnson's position cannot even be sustained within the context of the intelligent design movement that he is largely responsible for starting.

To say that scientific and religious statements are fundamentally different is not to say that "religious statements belong to the realm of faith while scientific statements to the realm of reason." Reason is not limited to science. Our scientific and theological understandings must inform each other if we are to be intellectually whole persons. We must strive toward an integrated Christian worldview that leaves no aspect of human activity or knowledge untouched. Maintaining clear definitions of different types of knowledge actually aids in their integration. The confusion of metaphysical naturalism with evolutionary theory actually inhibits the productive interaction between the sciences and Christian theology. It does so by injecting into a scientific theory a metaphysical worldview that is simply not a necessary component of the theory, and is in conflict with the very methods of science. In his response to Lamoureux, Johnson urges theistic evolutionists to get leading scientific organizations to "support a new statement unambiguously disavowing the mixing of scientific and religious claims." That is precisely what we have been working toward. Ironically, Johnson's own insistence that evolution is inseparably tied to metaphysical naturalism, and his efforts to have divine action incorporated into scientific description, are directly in conflict with that stated goal.

Johnson has criticized evolutionary theory by referring to specific observable data. He has devoted much of his argument over the years to casting doubt on the observational support for evolutionary theory. As such, he has made scientific claims that can only be refuted by discussing the available evidence. He must then be willing to accept criticisms of his scientific statements and respond to them. That is the nature of scientific discourse. Instead he states in his response to Lamoureux, "I do not think

it worthwhile to discuss detailed evidentiary questions with Denis Lamoureux, *or with other persons who take the position I call theistic naturalism,
whatever they choose to call it"* (my italics).[7] He thus effectively refuses to
discuss the evidence that he uses to support his position with any advocate of theistic evolution whose philosophy of science or theology of
God's creative action he finds objectionable. This insulation from criticism is not conducive to dialogue, or to the search for truth. For similar
reasons, Johnson fails to engage the depth of theological thought present
among his evangelical Christian brothers and sisters who accept evolution as the best scientific description of the history of life.

Johnson accuses Lamoureux of basing his arguments on appeals to
authority. However, he fails to distinguish an appeal to authority from
an appeal to the evidence. For me, Johnson's qualifications are really not
the issue. People from outside the scientific disciplines have historically
made some very important contributions to scientific thought. But he has
made demonstrably incorrect statements about both evolutionary theory
and the evidence upon which it is based. Attempting to point out those
errors is not an appeal to authority.

In *Darwin on Trial* Johnson spends considerable effort attempting to
cast doubt on the ability of the fossil record to support common descent.[8]
My acceptance of common descent is based on the overwhelming force
of the evidence. The fossil record is the evidence of which I am particularly aware, since that is my area of professional training and research.
There are numerous examples of fossils with transitional morphologies
crossing every taxonomic category from species to phyla. In the fossil
record it is found that representatives of different higher-level taxa (e.g.,
families, orders, classes) become more 'primitive,' that is, have fewer derived characteristics and appear more like the primitive members of
other closely related taxa as one moves back in time. As a result, species
that lived nearer the presumed branching points become increasingly
difficult to place in a higher taxon. Similarly, for lineages with a good
fossil record, the appearance of a new higher taxon is associated with the
occurrence of species whose taxonomic identity is uncertain or converges
closely on that of the new higher taxon. Such patterns are found repeatedly by paleontologists.[9]

---

[7] See page 46 of this book.

[8] See chapters four and six in Phillip E. Johnson, *Darwin on Trial* (Downers Grove, Ill.:
InterVarsity Press, 1991).

[9] For an extended discussion of transitional forms in the vertebrate fossil record see the
American Scientific Affiliation web article by Keith B. Miller (November 25, 1997), "Tax-

Almost everyone is familiar with the fossil record of horses, the earliest known fossil horse being *Hyracotherium* ('*Eohippus*'). The fossil record of the rhinos, tapirs and titanotheres is also relatively good, and the earliest representatives of these perissodactyl families ('odd-toed ungulates') are all very similar to *Eohippus*.[10] Furthermore, the most primitive ungulates (hoofed mammals) were the condylarths, and they include forms very similar to the insectivores, carnivores, and even primates, such that some genera and families of the condylarths have been previously assigned to these other orders.[11] Going back still farther in the fossil record, certain groups of reptile species near the appearance of unquestioned mammals possess virtually all the characteristics of mammals. Similarly, some species occurring near the appearance of unquestioned reptiles have been at times placed within both the amphibian and reptile classes.[12] Thus, the farther you go back in the fossil record, the more difficult it is to place species in their 'correct' higher taxonomic group. The boundaries become blurred. This is precisely the expectation of evolutionary theory.

Of special interest in the history of life are the morphological transitions associated with major adaptive shifts from water to land, land to water, and land to air. These major changes in mode of life opened up tremendous new adaptive opportunities for animal life. In recent years, exciting fossil evidence has been uncovered for several of these transitions. Lamoureux has already addressed the new discoveries relevant to

---

onomy, Transitional Forms, and the Fossil Record" at <http://asa.calvin.edu/ASA/resources/Miller.html>. A copy of this article can also be obtained by contacting the author.

[10] For detailed descriptions and discussion of these fossils, see B. J. McFadden, *Fossil Horses: Systematics, Paleobiology, and Evolution of the Family Equidae* (Cambridge: Cambridge University Press, 1992), 369; D. R. Prothero and R. M. Schoch (eds.), *The Evolution of the Perissodactyls* (New York: Oxford University Press, 1989).

[11] A. S. Romer, *Vertebrate Paleontology* (Chicago: University of Chicago Press, 1966), 468; R. L. Carroll, *Vertebrate Paleontology and Evolution* (New York: Freeman, 1988.), 698

[12] J. A. Hopson, "Synapsid Evolution and the Radiation of Non-Eutherian Mammals," in *Major Features of Vertebrate Evolution, Short Courses in Paleontology*, no. 7, eds. D. R. Prothero and R. M. Schoch (Knoxville: Paleontological Society, 1994), 190–219. For an up-to-date review of the evidence for mammal origins see the following articles in H.-P. Schultze and L. Trueb (eds.), *Origins of the Higher Groups of Tetrapods: Controversy and Consensus* (Ithaca: Comstock, 1991)—J. A. Hopson, "Systematics of the Nonmammalian Synapsida and Implications for Patterns of Evolution in Synapsids," 635–693; N. Hotton, III, "The Nature and Diversity of Synapsids: Prologue to the Origin of Mammals," 598–634; and M. Desui, "On the Origin of Mammals," 570–597. For illustrations and discussion of the amphibian–reptile transition, see Carroll, *Vertebrate Paleontology* and M. J. Benton, "Amniote Phylogeny," in *Origins of the Higher Groups of Tetrapods*, 317–330.

the origin of whales from primitive ungulates. The transition from water to land was one of the most significant events in the history of life. Several new species of primitive amphibians have been recently described, and new fossil evidence from the previously known genus *Ichthyostega* has come to light. The limbs are now known to have had a very limited range of movement, and the animal was not as well-adapted for terrestrial locomotion as previously thought. It also turns out that these amphibians had seven to eight digits on their hands, rather than the five of all later tetrapods.[13] The lobe-finned rhipidistian fishes are widely considered to have given rise to the amphibians. One group of rhipidistians, the panderichthyids, lived near the time of appearance of the first amphibians. Their lobed pectoral and pelvic fins have bones that compare with the limb bones of tetrapods. These fishes had flattened skulls very similar to that of the earliest amphibians, anal and dorsal fins were absent, and the tail was like that of *Ichthyostega*. The first known skull of a panderichthyid was in fact initially considered to be that of an amphibian.[14] This again illustrates the taxonomic problems encountered during the appearance and early radiation of a new taxon.

Evidence for the relationship of birds to dinosaurs has been growing at a spectacular rate, particularly in the last few years. Well over 20 shared characteristics have now been identified between *Archaeopteryx* and a certain group of theropod dinosaurs. Among these are a toothed skull, a wishbone, a theropod-like pelvis, the close similarities of the bones of the forelimbs including a swivel wrist joint, and the similarity of the hind limbs and feet with the presence of a reversed first toe.[15] The similarities of *Archaeopteryx* to theropod dinosaurs such as *Velociraptor* and *Deinonychus* are especially strong, and a newly discovered dinosaur has features of the limbs and pelvis that are the most bird-like yet known.[16] Spectacular new fossil finds in China have revolutionized our understanding of bird origins. Two new fossils, *Protoarchaeopteryx* and

---

[13] P. E. Ahlberg and A. R. Milner, "The Origin and Early Diversification of Tetrapods," *Nature* 368 (1994): 507–514.

[14] See H.-P. Schultze, "A Comparison of Controversial Hypotheses on the Origin of Tetrapods," 26–67; and E. Vorobyeva and H.-P. Schultze, "Description and Systematics of Panderichthyid Fishes with Comments on Their Relationship to Tetrapods," 68–109, both in *Origins of the Higher Groups of Tetrapods*.

[15] M. K. Hecht et al. (eds.), *The Beginnings of Birds: Proceeding of the International Archaeopteryx Conference, Eichstatt, 1984* (Eichstatt: Bronner & Daentler, 1985); J. H. Ostrum, "On the Origin of Birds and of Avian Flight," in *Major Features of Vertebrate Evolution*, 160–177.

[16] F. E. Novas and P. F. Puerta, "New Evidence Concerning Avian Origins From the Late Cretaceous of Patagonia," *Nature* 387 (1997): 390–392.

the short-armed *Caudipteryx*, are more primitive than *Archaeopteryx*, have symmetrical rather than asymmetrical feathers, and would have been flightless. These specimens blur the boundaries between dinosaur and bird. Also from China is the beautifully preserved fossil of a theropod dinosaur with a covering of downy-like filaments indicating the presence of feather-like coverings on some dinosaurs![17] Finally, in the last several years the discovery of new fossil birds from the Cretaceous has led to the erection of a whole new subclass of primitive birds. This new group includes several fossil specimens previously identified as theropod dinosaurs![18]

Johnson states in his response to Lamoureux that "I doubt that the common ancestry thesis is true, at least at the higher levels (phyla) of the taxonomic hierarchy."[19] However, even here, recent discoveries are closing the gaps. Much has been made of the rapid diversification of animals across the Precambrian/Cambrian boundary. Despite popular claims to the contrary, metazoans (multicelled animals) do appear as fossils before the beginning of the Cambrian. Several living phyla (sponges, coelenterates, molluscs, echinoderms, as well as several worm phyla) are now believed to be represented by the latest Precambrian fossils. Probably most astonishing is the discovery of fossilized embryos of metazoans in the late Precambrian that antedate the appearance of any known body fossils. These embryos appear to represent cnidarians (a group of coelenterates) and worm-like bilaterians.[20] Also, with major new discoveries in the last decade, the many unusual early Cambrian fossils have been recognized as primitive members of living phyla, or have been placed into new taxonomic groups with body plans that bear similarities to more than one living phylum.[21] Important new discoveries of exception-

---

[17] For beautiful colour photographs of these new specimens see Jennifer Ackerman, "Dinosaurs Take Wing," *National Geographic* 194, no. 1 (1998): 74–99.

[18] K. Padian and L. M. Chiappe, "The Origin and Early Evolution of Birds," *Biological Reviews* 73 (1998): 1–42; L. M. Chiappe, "The First 85 Million Years of Avian Evolution," *Nature* 378 (1995): 349–355.

[19] See page 43.

[20] M. A. Fedonkin and B. M. Waggoner, "The Late Precambrian Fossil *Kimberella* Is A Mollusc-like Bilaterian Organism," *Nature* 388 (1997): 868–871; C.-W. Li, J.-Y. Chen, and T-E. Hua, "Precambrian Sponges with Cellular Structures," *Science* 279 (1988): 879–882; R. A. Kerr, "Pushing Back the Origins of Animals," *Science* 279 (1998): 803–804; S. Bengtson and Y. Zhao, "Fossilized Metazoan Embryos from the Earliest Cambrian," *Science* 277 (1997): 1645–1648; S. Xiao, Y. Zhang, and A. H. Knoll, "Three Dimensional Preservation of Algae and Animal Embryos in a Neoproterozoic Phosphorite," *Nature* 391 (1998): 553–558.

[21] For a fuller discussion see Keith B. Miller, "The Precambrian to Cambrian Fossil Record and Transitional Forms." *Perspectives on Science and Christian Faith*, 49 (1997): 264–268.

ally well-preserved fossils of soft-bodied organisms in China, and the redescription of previously known specimens has resulted in the recognition of a diverse and widespread group of organisms called lobopods.[22] These caterpillar-like organisms walked on fleshy legs and bore plate-like or spine-like mineralized structures on their backs. The Cambrian lobopods occupy a transitional morphological position between several living phyla. The oldest known lobopod bears similarities to both palaeoscolecid worms and to living onychophorans and tardigrads. Furthermore, lobopods also have morphological features in common with the arthropods.[23] Another very important group of Early Cambrian fossils is represented by a wide variety of tiny cap-shaped and scale-like skeletal elements. It is now known that many of these belonged to slug-like animals that bore these hollow mineralized structures like an armour. These organisms are mosaics of phylum-level characteristics, and their taxonomic affinity is a matter of present debate. Some bear strong resemblances to the molluscs and others to the polychaete annelid worms.[24] The taxonomic confusion associated with these scale-bearing slug-like animals and with the lobopods is consistent with their stratigraphic position at the base of the Cambrian metazoan radiation.

The surprising number of spectacular 'gap-filling' fossil discoveries made just since the 1991 publication of Johnson's *Darwin on Trial* reveals how false is his assessment that "the fossil record today on the whole

---

[22] L. Ramsköld and H. Xianguang, "New Early Cambrian Animal and Onychophoran Affinities of Enigmatic Metazoans," *Nature* 351 (1991): 225–228; Jun-yuan Chen and B.-D. Erdtmann, "Lower Cambrian Fossil Lagerstatte from Chengjiang, Yunnan, China: Insights for Reconstructing Early Metazoan Life," in *The Early Metazoa and the Significance of Problematic Taxa*, eds. A. M. Simonetta and S. Conway Morris (Cambridge: Cambridge University Press, 1991): 57–76; L. Ramsköld, "Homologies in Cambrian Onychophora," *Lethaia* 25 (1992): 443–460.

[23] J. Dzik and G. Krumbiegel, "The Oldest 'Onychophoran' *Xenusion*: A Link Connecting Phyla?" *Lethaia* 22 (1989): 169–181; J. Dzik, "Early Metazoan Evolution and the Meaning of Its Fossil Record," in *Evolutionary Biology*, eds. M. K. Hecht, et al., 27 (1993): 339–386; Jun-yuan Chen, L. Ramsköld, and Gui-ging Zhou, "Evidence of Monophyly and Arthropod Affinity of Cambrian Giant Predators," *Nature* 264 (1994): 1304–1308; G. E. Budd, "The Morphology of *Opabinia Regalis* and the Reconstruction of the Arthropod Stem-Group," *Lethaia* 29 (1996): 1–14.

[24] J. Dzik, "Early Metazoan Evolution and the Meaning of Its Fossil Record," in *Evolutionary Biology* eds. M. K. Hecht, et al., 27 (1993): 339–386; S. Bengston, "The Cap-shaped Cambrian Fossil *Maikhanella* and the Relationship between Coeloscleritophorans and the Molluscs," *Lethaia* 25 (1992): 401–420; N. J. Butterfield, "A Reassessment of the Enigmatic Burgess Shale Fossil *Wiwaxia Corrugata* (Matthew) and Its Relationship to the Polychaete *Canadia Spinosa* Walcott," *Paleobiology* 16 (1990): 287–303.

looks very much as it did in 1859."[25] Seeing the history of life unfolding with each new discovery is exciting to me. God has given us the ability to see into the past and watch his creative work unfold. To do so is for me a very worshipful experience, and it has greatly broadened my perception of God's power and unfathomable wisdom. It has also informed my understanding of the meaning of suffering and the cross, and given new perspective to our call to be God's image bearers to the rest of creation. I encourage the evangelical church to look at the evolutionary record not as an obstacle to faith or challenge to evangelism, but as a cause for renewed contemplation of the truths revealed in Scripture.

---

[25] Johnson, *Darwin on Trial*, 50.

On Being and Becoming: Conflation and Confusion of the 'Science' of Evolution

Michael W. Caldwell

In *Defeating Darwinism*, the catalyst for the debate between Phillip Johnson and Denis Lamoureux, Johnson details his ideas on how young people, "need to protect themselves against the indoctrination in naturalism that so often accompanies education. Textbooks and other educational materials today take evolutionary naturalism for granted, and thus assume the wrong answer to the most important question we face: Is there a God who created us and cares about what we do?"[1]

Like Lamoureux, as a scientist and educator I must respond to accusations that the curriculum I support indoctrinates young minds and avoids important questions. In responding to such an accusation my responses are, admittedly, rather limited. If I agree that Johnson's charges are accurate then this commentary on the Johnson–Lamoureux debate can be quite short leaving all of Johnson's writings to stand as scripture to his testimony. However, if I state that he is incorrect, then the burden of argument is upon me to find the flaws in Johnson's rhetoric and present a case for a more balanced form of reason.

I have chosen to take the second stance. In the following commentary my arguments focus on Dr. Johnson's veracity and legitimacy as a critic of science, science education, and the philosophy of science, in the context of his three books and his debate with Denis Lamoureux.

Science: On the Apple and How We 'Know'

The thesis for this portion of my commentary is that Dr. Johnson's grasp of the factual content of biology, paleontology, and geology is, at best, poor. In this context, my definition and use of *factual* or a *fact* is categori-

---

[1] Johnson, Phillip E. *Defeating Darwinism by Opening Minds* (Downers Grove, Ill.: Inter-Varsity Press, 1997), 10.

cal—in other words, if you hold a red apple in your hand and are asked to name and describe what you see, you will tell me you have a red apple in your hand. This form of strict object definition requires that the observer have as categories in his or her own mind a notion of 'red' and a notion of 'apple'. Red is hard to describe beyond the statement 'red', unless contrasted with 'yellow', but this only serves to define red by a category of 'not red'.

'Apple' on the other hand, comes defined in the observer's mind with a variety of shape categories. In biological jargon, these shape categories are the morphology of the object being described. For an apple this involves a rounded to heart-shaped outline, a thin woody stem set down into the top of the fruit, a dry leafy circle on the bottom of the fruit that represents what is left of the flower, and so the list goes on. A very long time ago Aristotle, the ancient Greek philosopher, labelled such categories as typological, and categorical thinking as typology.

Bearing this description of the 'fact' of the apple in mind, it is possible to proceed to examine the 'facts' of biology, paleontology, and geology: the Life and Earth sciences. As a basic method of seeing and thinking about the natural world, these sciences are typological; they collect things, give them names, and provide categories in which to classify the collected things, or objects. In brief overview, the facts of the Earth and Life sciences are the tens of millions of fossil specimens representing everything from bacteria to humans. These facts are the two to three million currently described species of living things, the suspected five to eight million more, the mechanisms of inheritance, the food webs, the structures of ecosystems, and so on, and so on. Now, it should seem simple to keep all these facts straight in our minds, and to communicate them, if we can only arrange them and organize them like the *Encyclopaedia Britannica* or perhaps the *World Book Encyclopedia*. Is this possible? The answer is yes, it can be done.

In the typological sense we can organize all these facts. However, in science we are also in the business of generating ideas that connect the 'facts'. Herein lies the dilemma! As I have shown with the apple, facts are categorical. In evolutionary science terms I do not make reference to a 'fact' as a transitory form supporting particular evolutionary scenarios, nor is a 'fact' considered as 'evidence' supporting hypotheses of evolutionary relationship. Transitory forms are objects that are hypothesized to connect two other objects in time and space. So, let us return to the 'apple' to examine this problem of 'making connections'.

122

If I ask the observer another question, say, "What is the thin woody stem at the top of the apple for?" he or she will likely reply, "The woody stem is the part that once served to fix the apple to the tree."

Suppose I ask the observer to cut the apple in half. What will be found? A bunch of seeds in the centre of the apple, surrounded a by thin membranous cover, all of which is surrounded by a thick, crispy fruit. Suppose I ask the observer what the small brown objects in the centre of the apple might be? I will likely be told they are seeds. And if I ask what seeds are for, I will be told that if the seeds fall into the ground they will sprout and become a tree.

What has happened here? From our Aristotelian category of 'red apple', we have somehow leaped to two absolutely non-categorical statements regarding our red apple: (1) The apple was once attached to the tree; (2) The seeds will become a tree. These are statements and conclusions we arrive at everyday as we make our way in the world. They seem absolutely normal—and they are—but they are also fundamentally different from categorical statements. We have now related things that we have not seen, both in the past and in the future, to the object we hold in hands. In science, the statement that the apple was once attached to a tree is referred to as a reconstructive or historical hypothesis of relation. That the seeds will become trees is a predictive statement—plant the seeds and observe them grow into a tree.

The seeds, the apple, the stem, and the tree are facts. The relationship between them is hypothetical until tested: planting the seed and observing the growth of a tree leads to a theory of plant development. In science, this difference between the facts and the hypotheses or theories that form the links between facts is a classic dichotomy, the dichotomy of 'pattern' and 'process'. For Aristotle, who three thousand years ago saw the same problems of fact and theory, the problem was the dilemma of 'being' and 'becoming'. The apple as a category in Aristotle's typological world was in a state of 'being'. At issue for Aristotle, as it is today in philosophy and the sciences, is the problem of 'becoming'. We see the apple, yet oddly we understand the process that made the apple and the process that awaits the apple. Like 'fact and theory' and 'pattern and process', 'being and becoming' are states of things and the ways we make connections between them. To be accurate critics of a system of seeing and thinking about the world we must know how to keep them separate in our minds.

Rephrased, the basic question I posed at the beginning was, "Is Dr. Johnson's demonstrated knowledge of the Earth and life sciences sufficient to allow him to teach the biology classes of the students whose education he wishes to save?" My line of argumentation will show that it is not, and more to the point, this insufficiency has led to his conflation and confusion of fact with theory, pattern with process, or to turn yet another very ancient phrase, being with becoming.

Facts, Patterns, and the Difficulty of Being

Suppose we now talk about a fossil. What are the factual or typological statements that might describe a fossil? In one of his earlier books, Dr. Johnson speaks critically, and with the appearance of authority, of the 140-million-year-old fossil *Archaeopteryx*, stating that it is a bird, not a good transitory form, and possibly, but not absolutely, an ancestor of birds. [2] Johnson cites several paleontologists on their definition of what constitutes a 'possible ancestor,' and uses these statements to critique the transitory form or missing-link hypothesis. Johnson privately confides that *Archaeopteryx* could be nothing more than an odd variant of bird. My questions then are legion and harken back to the apple.

Has Dr. Johnson ever seen *Archaeopteryx*? If given a bony skeleton of a chicken, could Dr. Johnson name a single bone? What is Dr. Johnson's definition of a bird? How does he recognize 'birdness' in both chickens and *Archaeopteryx*? Most bird fossils have neither feathers nor soft tissues preserved (e.g., eggs), but do have teeth — real teeth, not serrations on the beak. Modern birds do not have teeth but can be recognized as birds because only modern birds have feathers and lay eggs. What criteria constitute birdness at this level for Dr. Johnson? What criteria will he use to differentiate species? And, how does Dr. Johnson treat the features of a fossil like *Archaeopteryx* where those features break the typological constraints he has formulated above and show a striking number of dinosaur, not bird, features? If dinosaur characters are bird characters for Johnson, then wings must not be a bird feature, claws must be allowed, and long-necked sauropods dinosaurs must be gigantic lumbering birds. The dilemma goes on ad infinitum.

Remembering our discussion of the red apple, what then is the basis for Dr. Johnson's dismissal of *Archaeopteryx*? It is a conflation and confusion of factual and theoretical statements. In other words he assumes definitions where none exist and, even worse, he is incapable of defining

---

[2] Johnson, Phillip E. *Darwin on Trial* (Downers Grove, Ill.: InterVarsity Press, 1991), 80–81.

them himself, and then confuses typological and transformational descriptions as if they were equivalent statements. For example, if a chicken is our apple, and *Archaeopteryx* is an apple pear, Dr. Johnson has concluded that the apple pear is only an apple (without any clear knowledge of categories), and then conflates his categorical confusion with unsupported process hypotheses by stating that the apple pear is not an example of a genetic cross between an apple and a pear. Because he holds an idea of 'fruit' in his mind, an idea that is both typological and transformational, his fruit tree can be expanded to produce oranges, tomatoes, grapes, apples, and apple pears. Fruit is fruit and the tree of origin is of no consequence.

Is *Archaeopteryx* the only example of such errors? No, it is not. Dr. Johnson, in *Darwin on Trial*, dismisses the morphology of a very odd mammal, the living duck-billed platypus, where he says, while speaking of *Archaeopteryx*:

> of those odd variants, like the contemporary duck-billed platypus, that have features resembling those of another class but are not transitional intermediates in the Darwinian sense.[3]

As a putative enlightener of minds, he raises more questions without answering a single one. What is a transitional intermediate, and what is such an intermediate in the Darwinian sense? And further, how does Johnson 'know' that the playtpus is not a Darwinian intermediate? Does he provide a means of testing putative intermediates for both 'intermediateness' and 'Darwinian intermediateness,' so that upon testing they are shown absolutely to be outside of those categories? If speaking to students of any age, how would he, wearing the mantle of a typologist, define or classify the duck-billed platypus, as we would our apple, given the following characteristics: (1) hair or fur; (2) warm-blooded, though just barely; (3) lays eggs; (4) lactates, not from a mammary gland, but rather from a small pore at the base of a hair down which dribbles a fatty, liquid substance; (5) possesses a hip and shoulder-girdle skeleton not at all like that of a modern mammal. This is not a call for a classification of transformation, I simply challenge him to define his odd variants. Odd variants must fit somehow in his classification scheme but he does not explain how. I simply lay out the challenge to provide a typological–Aristotelian category for the platypus, in the same way that I challenge him to explain how *Archaeopteryx* is a bird. Once I am convinced he can

---

[3] Ibid., 80.

define birdness and mammalness, then and only then can I concede his inclusion of *Archaeopteryx* within birds, and the platypus within mammals, and further, why neither animal represents an intermediate to anything else!

I can go on concerning Johnson's factual knowledge of science, but I think this is an issue that Lamoureux has undertaken as well — it is also an issue that fades away in the light of an even more basic misunderstanding of the theoretical systems of science.

Theories, Processes, and the Difficulty of Becoming

That a red apple can be understood in our minds as both an object and an incipient something means that the conceptualizations of processes ('becoming') are natural psychological frameworks we produce for organizing our perceptions of the world around us. We make connections. However, if we do not understand the objects being connected, the theoretical systems we construct to connect those objects will be confused.

This kind of confusion surrounds many scientific ideas; evolution is but one of many. What is meant by the word? Is it a theory, an idea, a collection of theories? What theoretical systems support the word or idea 'evolution'?

I believe from my readings of Dr. Johnson's writings that he and I would agree on several points: (1) the universe is old; (2) the earth is old but younger than the universe; (3) the age of the earth can be determined using the tools of geology; (4) rocks at the bottom of a stratigraphic sequence, assigned to the Precambrian system of rock-time units (old rocks), do not contain fossils of humans; (5) Precambrian rocks contain fossils of colonial or isolated bacterial cells; (6) rocks higher in the sequence contain trilobites and bacteria; (7) trilobites are not alive today; (8) dinosaurs are not found in rocks with trilobites; (9) dinosaurs in older rocks are different from dinosaurs in younger rocks. In short, apart from explanatory mechanisms (theoretical systems), Dr. Johnson and I would agree that organisms change over time as recorded in a sequence of rock units.

Long before Darwin, when time was first recognized as possibly being very 'deep', and when the above examples of organismal variation through time were first recognized, this change over time was referred to as 'evolution.' This is all that evolution refers to — change over time. No mechanism was ever postulated, nor for that matter was it required. Even a superficial reading of pre-Darwin treatises in theoretical biology and geology shows that the idea predates Darwin.[4] If the idea of evolution was not Darwin's, then we must ask, what ideas are Darwin's, what did he say about them, and just what is evolution?

Darwin constructed the theory of natural selection and recognized a pattern of change that is now referred to as gradualism.[5] Darwin's natural selection is a theoretical system that describes how the natural environment selects adaptable organisms, and how that adaptability affects an individual's reproductive success and the subsequent constitution of future gene pools. The modern expression for Darwin's theory of natural selection is 'differential reproductive success.' Darwin observed breeding and inheritance among domesticated animals and plants and noted how specific features of an animal could be 'selected' to appear in subsequent generations. In the natural world he observed similar patterns of variation among animals as they responded to their particular environments.

Natural selection is not synonymous with gradualism. Gradualism is a pattern that Darwin recognized as to how slowly and gradually features change over reproductive generations; natural selection is a process that Darwin hypothesized would select from among natural variants occurring within a species. Gradual change, or what is now referred to as microevolution was the pattern of change Darwin saw when he looked at the variation between generations of sexually reproducing organisms.

Today, as it was for Darwin, the fossil record does not and cannot test natural selection. Natural selection, as a theory of process, requires that morphology, as transformed by inheritance between generations, be observable and measurable relative to some kind of external perturbation. Darwin recognized that the pattern of change shown by organisms found in the fossil record did not fit well with his observed pattern of gradual change between generations. It was very difficult if not impossi-

---

[4] Charles Bonnet (1720–1793), an eighteenth-century biologist from Geneva, first coined the term *évolution*, and at the same time his synonyms of *révolution* and *métamorphose*, in the context of understanding development, nascent ideas of extinctions, and his construction of a classification scheme for living things and how those living things might transform over time. Subsequent developments in the usage of the word 'evolution' to its present form involved a large number of eighteenth- and nineteenth-century thinkers such as Lamarck and, eventually, Darwin, employing and being influenced by a great number of ancient and, at the time, modern philosophies. A superb review is given by O. Rieppel in his *Fundamentals of Comparative Biology* (Basel: Birkhäuser Verlag, 1988).

[5] Carroll, R. L. *Patterns and Processes of Vertebrate Evolution*. Cambridge Paleobiology Series, (Cambridge: Cambridge University Press, 1997), 19–33. Carroll provides a good overview of Darwin's notions of gradual changes accumulating over time—a pattern now referred to as gradualism or, as coined by Eldredge and Gould, "phyletic gradualism."

ble to find fossil assemblages that would preserve as detailed a record of change between generations as he could record by observing the effects of breeding in pigeons. It was not until 1972 that Gould and Eldredge[6] found a way to explain the pattern of change seen in the fossil record: punctuated equilibrium or macroevolution are terms given to process explanations of the fossil record pattern that shows large and rather episodic jumps in morphology between species.

The theory of natural selection is predictive. Tests of the theory can be deduced from observation of the adaptations of organisms to their environments. Natural selection cannot test hypotheses of historical organismal change over time, but it can serve as a predictive test of future change. In contrast, the fossil record serves as a historical test of change over time, not as a test of natural selection, and not as a predictor of future organismal variation. The fossil record can be held up as an example of the success of certain innovations such as the evolution of the ear through the therapsid to mammal transition, but even the success of hearing systems does not test natural selection. It simply stands as a mute inductive witness to Selection. However, the fossil record can serve as a predictive test of changing morphology through time, and on this account it has never been falsified! Not even the argument of a lineage or taxon that can be identified as morphologically static for millions of years of rock-time falsifies the prediction; the static morphology is still different from those organisms found in older rocks. The predictive question need not reflect 'progressive change', 'improvements', etc., for these are theory laden and require ad hoc or special hypotheses as prior structures. No, the predictive test only requires that some character (typological category) be different between forms occurring at different levels in the rock record.

Likewise, tests of deductive hypotheses, that through falsification could render the theory of natural selection impotent, have never been demonstrated. Johnson cites the example of the peppered moth of Industrial and Post-industrial Revolution England as non-evidence of Darwinism and a faint but feeble grab at empiricism. From the above discussion it is clear he does not understand natural selection. The peppered moth example, and hundreds of thousands of similar examples, published monthly in science journals, are excellent examples of genetic

<hr>

[6] Eldredge, N. J., and Gould, S. J., "Punctuate Equilibria: An Alternative to Phyletic Gradualism," in *Models in Paleobiology*, ed. T. J. M. Schopf (Freeman Cooper, San Francisco, 1972), 82–115.

variation existing in the gene pool of species that, when selected for by environmental factors, produce more or less adaptable morphological variants. This is natural selection and it works exactly as predicted. The poor fit is not between theory and observation, but between Dr. Johnson's understanding of fact and theory.

Johnson simply does not know, as Lamoureux correctly points out, that there is no such formal structure as Darwinism. He does not know that natural selection is Darwin's only contribution to what is collectively referred to as 'post-Darwinian evolutionary theory.' He clearly does not know that the modern theoretical package is a collection of process theories that includes natural selection as well as a number of genetic processes under the umbrella of population genetics. Gradualism and punctuated equilibrium are patterns, not processes of change through time; process remains elusive but why is this unreasonable? How has evolution science failed if, like medicine or physics, there remain intriguingly difficult theoretical problems for future generations to solve?

In summary, that Johnson confuses change over time with natural selection, gradualism, or punctuated equilibrium only means that he does not understand the nature of the factual information, the nature of the theoretical systems, nor the information sources that the disparate theoretical systems claim support from.

Education and Educators

Because Johnson confuses and conflates factual and theoretical systems in science he is not fit to be critic of science education. That he assumes he is qualified to criticize and then to exhort students simply means he has a very poorly conceived pedagogy, or philosophy of teaching.

Does an absence of factual information disallow an individual the right to, his or her opinion? The answer to this must be a definite no! However, what it does mean is that Johnson's motives as an educator must be seen as suspect when he attempts to diminish the importance of data sets and theories employed by people who do understand them. His motives become even more suspect when he takes the time, in his own writings, to disavow any responsibility to know the facts of the science he is roundly confused by and critical of:

> Such details as the difference between various vertebrate paleontology textbooks, and the precise identification of the candidate mammal ancestor for whales, are a sideshow, aimed at diverting the discussion

away from the main philosophical questions and into a morass of technical details.[7]

So, when the going gets tough in the factual world, Johnson's preference is a rhetorical sidestep to philosophical issues? So be it. However, before I examine Johnson's understanding of the philosophy of science, I have a few closing statements on the importance of teaching—an educational philosophy of my own.

As a professor of law at Berkeley, I would hazard a guess that Dr. Johnson does not exhort his students to ignore the morass of technical details in recent statutes from the State Legislature of California by categorizing them as "legal sideshows". This is shallow and empty rhetorical nonsense. That Dr. Johnson cannot juggle the bowling pins of science does not lessen the importance of juggling to a circus. As a university professor of biology and geology I would never presume to have special insight on the education of students of law. Education is a process of communication between those teaching and those learning. In my classes I present information. Some of that information is factual and some of it is theoretical. Students must learn the nexus of information that is created by theories and their supporting facts. I communicate, they study; I reinforce, redefine, and then, if all has gone well, they will revise the structure of the entire enterprise. In short, my task is to provide an intellectual milieu in which new theories will arise from the collection of new facts and reinterpretation of the old. Dr. Johnson presumes no such responsibility and gives no such right. For Johnson there is only one question of any importance, all others are either subservient or are superfluous and irrelevant along with their morass of supporting facts.

Philosophy

Someone once said, that what goes around comes around. Now is that time. In accord with Johnson, I will assume that the facts of science are a sideshow. Therefore, it is time to engage the issue of philosophical naturalism and the 'devious naturalistic assumptions' that underlie Johnson's version of evolutionary science.

Throughout his exchanges with Lamoureux, and throughout his books, Johnson continually belabours the point that evolutionary science is not a science but a thinly veiled platform for preaching a philosophy of metaphysical naturalism. It is clear from his criticisms that he sees this

---

[7] See page 47 of this book.

philosophical position, both as a worldview and as a system of assumptions in science, as inherently wrong or even delusional. Though an interesting debate on their own, moralities and the imperative to condemn are not the issue of my commentary; rather, I respond to just what metaphysical naturalism is, what the philosophy of science is, and pose the question, "Just who is free of metaphysical naturalism as a suite of assumptions used to understand and participate in the material world around us?" Or, in other words, are evolutionary scientists the only people who suffer from Johnson's supposed philosophical malady of naturalism and materialism?

Dr. Johnson wants us to accept a science that includes investigations and assumptions that search to prove the existence of God by reference to the material and natural world. Therefore, the next question is, what is Johnson's philosophy of science and of the material and natural world? How would Dr. Johnson see and describe the apple, and what form of hypotheses would he construct? Does he hold a theistic metaphysic to the exclusion of metaphysical naturalism, or does he hold a foundational naturalism as the root of his material philosophy of self, the world, and the apple? And, if there is as much confusion in his philosophy as there is in his knowledge of the factual and theoretical knowledge of science, can he any longer be accepted as a valid critic of any aspect of modern science?

The intellectual constructions of Karl Popper dominate the assumptions and method of modern science.[8] Popper's ideas developed as a contrast to the 'empirical' philosophy and methodology promoted by a group of intellectuals known as the Vienna Circle. The difference between Popper and the Vienna Circle was epistemic in nature, that is, their differences concerned knowledge and how we justify our claims for knowledge (similar to the apple we *see* versus the apple we *know*).

To use the jargon, the Vienna Circle claimed the reality of absolute knowledge, and set out to verify the existence of *a priori* truths derived from the construction of inductive hypotheses. The larger program of philosophical research using this method was known as positivism, and held that only sense-data informs the observer of the true nature of reality—if you can't touch it, like God, it can't be real or true!. Or, in plain English, they held that it was true that absolute truth existed and that it could be discovered by experiments that answered questions designed to be verified by those experiments. In short, it would be hard to prove the

---

[8] Popper, K. R. *The Logic of Scientific Discovery* (London: Routledge, 1959).

question wrong as it could be interpreted very broadly. This is similar to the dilemma Dr. Johnson creates by his undefined category of bird that can include *Archaeopteryx* and at the same time prove that it is not a transitory form.

In contrast, Popper argued that inductive logic was not equal to deductive logic in terms of its post-testing, or resultant, explanatory power. This was because, while deductive hypotheses could never be proven absolutely true nor absolutely false, deductive hypotheses could, by failing the tests they were submitted to, be shown to be relatively false. New data, new observations, and renewed tests would indicate the relative falseness of hypotheses; hypotheses passing the tests would be considered corroborated, but never absolutely true. This philosophy and methodology is known as falsificationism. The key to Popper's empiricism is his non-acceptance of absolute truth as a falsifiable epistemic (knowledge) framework. In other words, absolute truths could never be absolutely proven nor falsified and were therefore not suitable platforms on which to construct hypotheses.

Popper rejected inductive logic as a logical form equivalent to deductive logic, but did accept induction as a psychological phenomenon. At this level inductive thinking allows human beings to construct hypotheses, but these hypotheses result in universal statements that are non-falsifiable (e.g., God exists). Pursuing this further, Popper went to great pains to point out that metaphysical programs, in their search for absolute truths could never be tested empirically. He did not reject metaphysical research programs as useless, but rather simply excluded them from the world of empiricism, as he excluded inductivistic or positivistic approaches in science.

In *Darwin on Trial*, Johnson devotes chapter 12, "Science and pseudo-science," to a treatment of Popper.[9] This sounds good, except that Dr. Johnson applauds Popper's philosophical achievements only where it is safe to do so. In the middle of chapter 12 Johnson discusses Popper's exposé of logical positivism, and then gives the appearance of treating Popper on metaphysics. But it is only appearance. He highlights nothing of Popper's treatment of metaphysics as regards absolute truth, nor does he illustrate the truth-search of metaphysical research programs as comparable to positivism and the construction of non-falsifiable hypotheses.

---

[9] Johnson, Phillip E. *Darwin on Trial*, 2nd ed. (Downers Grove, Ill.: InterVarsity Press, 1993), 147–156.

Michael W. Caldwell

So why does Johnson obfuscate Popper? Johnson wants 'science with God' as a governing paradigm, yet he knows that metaphysical searches for knowledge are not empirical, they are positivistic, verificationist research programs. Johnson wants the cloak of empiricism. It is necessary and vital to his dependence on reason. Where does he accomplish this segue? In *Darwin on Trial* Johnson sidesteps from metaphysics, without using the word religion or Christian, to a sentence on Freud and Adler as examples of metaphysical pseudoscientists, to the following:

> Because of these complications, the falsifiability criterion *does not necessarily differentiate* natural science from other valuable forms of intellectual activity. Popper's contribution *was not to draw a boundary* around science, but to make some frequently overlooked points about intellectual integrity that are equally important for scientists and non-scientists [italics mine].[10]

Johnson's appeal and philosophy are clear. He avoids Popper on metaphysics by obfuscating metaphysics as a complication Popper encountered but did not feel was offensive. This is a clear misrepresentation of Popper[11] that allows Johnson to appear allied with Popper, and then to conclude, with the appearance of protection from the falsifiability criterion, that there is an absolute truth that can be searched for empirically: God. He holds that natural science and other valuable forms of intellectual activity can co-exist, or that naturalism and "other activities" (the search for the proof of God?) can be blended together. But Johnson's misrepresentation of Popper does not stop here, it continues with the second sentence of the above quotation where Johnson misleads the reader by trying to indicate that Popper did not create a boundary for science. From the first chapter of his most famous work on the philosophy of science, Popper writes:

---

[10] Johnson *Darwin on Trial*, 2nd ed., 151

[11] In *The Logic of Scientific Discovery* (chapter 1, section 4, 34–39), Popper sets out the problem of demarcation and its relationship to inductive vs. deductive logic. He discusses the positivistic approach and breaks down the arguments against metaphysics as highlighted by positivism. He does not defend metaphysics as a way to what he would later define as objective knowledge (see K. Popper, *Objective Knowledge: An Evolutionary Approach* [Oxford: Oxford University Press, 1972]), but rather attacked the arguments of positivism against metaphysics, and pointed out that both positivism and metaphysics generate "systems of meaningless pseudostatements" that obfuscate science (pg. 37, *The Logic of Scientific Discovery* [London: Routledge, 1959].)

> Finding an acceptable criterion of demarcation must be a crucial task for any epistemology which does not accept inductive logic. [12]

and,

> I still take it to be the first task of the logic of knowledge to put forward a concept of empirical science, in order to make linguistic usage, now somewhat uncertain, as definite as possible, and in order to draw a clear line of demarcation between science and metaphysical ideas — even though these ideas may have furthered the advance of science throughout history. [13]

So, finally, what does all of this mean and how does it relate to the above quote? First, Johnson is a positivist. He is a naive supporter of verificationist research programs, and a naturalist, and a materialist. He holds a fundamental naturalism at the root of his worldview and adds yet another philosophical assumption: A theistic metaphysic as explanans of a purposeless and directionless metaphysical naturalism. For Dr. Johnson, God can be proven to exist by verifying that existence through the direct evidence of every single object in the natural world. The problem is that there is no such proof outside of faith. The statement "God exists" can be verified as easily as "God does not exist;" however, in each case, the positive and negative existence of God can only be verified by 'evidence' interpreted before it is collected. This has been done. Popper argued against such inductive approaches, and they are now recognized as empirically bankrupt; Johnson's version of science is nothing but a positivistic research program seeking proof for God. Likewise, Johnson's Popper is nothing but a convenient rhetorical appeal to authority from a non-Popperian metaphysician who periodically and conveniently dons a shadowy Popperian cloak. There is no integrity in such scholarship.

In short, as with the 'being' and 'becoming' of things, the facts and theories of science, Dr. Johnson confuses, conflates, and misrepresents all of the issues and assumptions of the philosophy of science, without comprehending the ramifications for his own position.

Conclusion

In the final analysis, it is clear that Dr. Johnson knows nothing of the factual content and theoretical systems of the science of 1998 — this is an important deficit that cannot be overcome by rhetoric discounting such

---

[12] Popper, *The Logic of Scientific Discovery*, 27.
[13] Popper, *The Logic of Scientific Discovery*, 35.

information as a sideshow. Concerning his philosophy of education, his own approach seems quite clearly to support indoctrination—not the opening of minds. Again, information is information, it is not a sideshow. Because his own philosophy on and of science is flawed, there is nothing empirical or scientific about the science education for which Johnson claims to be an agent. Johnson's claim to embrace Popper's philosophy of science flies away as does his own inheritance of the wind when Popper is truly displayed against Johnson's inaccurate characterization. The tools Johnson puts forth as essential for opening minds are nothing but the tools of misinformation. Therefore, Johnson can in no way serve as a credible critic of Darwinism, evolutionary biology, paleontology, the philosophy of science, and most importantly, of science education.

And, lastly, as a postscript in defence of the intellectual man Charles Darwin, I must counter Johnson's criticism of Darwin for being an 'inductivist' with the following rhetorical question, "What else could he be?" Every single thinker at the time (including those before Darwin, such as Galileo!) lived long before Karl Popper formalized deductive logic as the foundation of empirical science. Therefore, Johnson's hindsight criticism of Darwin makes about as much sense as using a New Testament exegesis to condemn Moses for worshipping God through Judaism.

The Intelligent Design Movement, Evangelical Scientists, and the Future of Biology

Jonathan Wells

When I was growing up in the northeastern United States, I loved ice skating with my friends on frozen ponds in the wintertime. We would grow wary, however, as spring approached and the ice became progressively weaker. Although it may still look solid on the surface, spring ice cannot be trusted.

The Darwinian paradigm in the late 1990s is like spring ice. Although its defenders continue to insist that it is supported by overwhelming evidence, it is quietly being questioned, revised — even abandoned — by a growing number of biologists. The Darwinian paradigm is in serious trouble, of the kind that matters most in science: it doesn't fit the evidence.

Of course, Darwin's theory works very well at some levels. According to Darwin, descent with modification is the result of natural selection acting on small variations. We can observe this within species (as in the breeding of domestic animals, or the emergence of antibiotic-resistant strains of bacteria), and we can infer it from adaptive differences among similar species in the wild (as in beak-shape differences among finches on the Galapagos Islands). Darwin believed that the same process, given enough time, could account for the large-scale evolution of all living species from one or a few common ancestors. Here, however, Darwin's theory runs afoul of the evidence.

First, Darwin maintained that major differences evolve from minor ones. Yet the fossil record shows that all of the major animal groups appeared at approximately the same time, without any fossil evidence that they diverged from a common ancestor. These original groups have since diversified into many subgroups, so the major differences among animals appeared before the minor ones. Paleontologists James Valentine and Douglas Erwin call this a "seeming paradox," since in this respect Darwin's theory "does not accord with the primary evidence."[1]

Second, in order for natural selection to produce evolution, a population must include suitable variations. No one doubts that natural populations include variations, but are those variations the kind that can lead to large-scale evolution? Modern research on the genetic basis of adaptation suggests that they are not. Geneticist John McDonald considers this "a great Darwinian paradox," since those genes "that are obviously variable within populations do not seem to lie at the basis of many major adaptive changes," while those that "seemingly do constitute the foundation of many, if not most, major adaptive changes apparently are not variable within natural populations."[2] In other words, the variations we see are not the ones Darwin's theory needs, and the ones it needs we don't see.

Third, the modern Darwinian view assumes that organisms develop according to genetic programs; presumably, mutations in those programs lead to evolution. When molecular biologists recently discovered that animals as different as mice, flies, and worms have similar developmental genes, some biologists saw this as confirmation that these animals inherited their genetic programs from a common ancestor. Other biologists, however, saw a profound puzzle: If organisms develop according to genetic programs, then how can similar genetic programs make a mouse, a fly, and a worm? If it is not their genetic programs that make these animals so different, then what is it? Embryologists John Gerhart and Marc Kirschner call this yet another "paradox," and suggest that evolutionary biologists have been "looking in the wrong place."[3]

As a biologist, I am fascinated by these paradoxes. I am also fascinated by the reluctance of many scientists to acknowledge them. Unfortunately, some people have anti-religious motives for defending Darwinism; as Richard Dawkins put it, "Darwin made it possible to be an intellectually fulfilled atheist."[4] This anti-religious dimension may ex-

---

[1] James W. Valentine and Douglas H. Erwin, "Interpreting Great Developmental Experiments: The Fossil Record," in *Development as an Evolutionary Process*, eds. Rudolf A. Raff and Elizabeth C. Raff (New York: Alan R. Liss, 1987), 96–97.

[2] John F. McDonald, "The Molecular Basis of Adaptation: A Critical Review of Relevant Ideas and Observations," *Annual Review of Ecology and Systematics* 14 (1983): 93.

[3] John Gerhart and Marc Kirschner, *Cells, Embryos, and Evolution* (Malden, Mass.: Blackwell Science, 1997), 129, 140.

[4] Richard Dawkins, *The Blind Watchmaker* (London: Penguin; New York: W. W. Norton, 1986), 6.

plain why, in the face of mounting empirical difficulties, the Darwinian paradigm has hardened into a sort of secular orthodoxy.

But more and more biologists, even non-religious ones, are looking for alternatives to Darwinian evolution. For example, Brian Goodwin writes: "Darwin's assumption that the tree of life is a consequence of the gradual accumulation of small hereditary differences appears to be without significant support. Some other process is responsible for the emergent properties of life, those distinctive features that separate one group of organisms from another. . . . Clearly something is missing from biology."[5] I am inclined to think that minor adjustments to Darwin's theory will not suffice, and that we must step outside the current paradigm and take a fresh new look.

One group which is doing just that is the rapidly growing intelligent design movement. Defenders of Darwinism criticize intelligent design for being contrary to the basic principles of science, but they are surely mistaken: most western scientists before Darwin, including the founders of modern paleontology, genetics, and embryology, regarded living things as designed. It is Darwinism, not science itself, that excludes design, and future progress in biology may depend on rethinking that exclusion.

Under the circumstances, it saddens me that so many evangelical Christians side with Darwinian critics of the intelligent design movement. Although there are notable exceptions, a surprising number of evangelical scientists and scholars seem overly impressed by their secular colleagues and timidly defer to Darwinian claims instead of critically examining them. But evangelicals have no business defending a secular orthodoxy, especially in the face of mounting counter-evidence. Instead, they should be encouraging the sort of bold new approach championed by the intelligent design movement.

Whatever the future of science may be, it will certainly be different from the present. Biology needs creative thinkers who can lay a foundation for the future, not apologists for a crumbling paradigm. Spring is coming, and it's time to get off the ice!

---

[5] Brian Goodwin, *How the Leopard Changed Its Spots* (New York: Charles Scribner's Sons, 1994), viii–ix.

Comments on
Special Creationism

Michael J. Denton

B ecause I have not read Phillip Johnson's latest book *Defeating Darwinism* I am unable to comment specifically on the arguments he presents there. However I have read *Darwin on Trial* and *Reason in the Balance* and have some grasp of his position.[1] I have also read the article about Johnson in December's issue of *Christianity Today* entitled 'The Making of a Revolution.'[2] I have also gleaned something of Johnson's views from personal communication with Johnson and from Denis Lamoureux's review of *Defeating Darwinism* entitled "Evangelicals Inheriting the Wind: The Phillip E. Johnson Phenomenon," from Johnson's rebuttal and Lamoureux's reply to his rebuttal.[3] I have not a great deal to add to Lamoureux's paper. But I would like to stress, as does Lamoureux, the importance of distinguishing between the term 'evolution' and 'Darwinism', and to make a few minor additional points about the way Johnson makes use of the 'gaps' to argue for divine intervention, say something about the facts of geographical distribution, and comment on the concept of intelligent design and on Johnson's philosophical naturalism.

The Terms 'Darwinism' and 'Evolution'

As Lamoureux rightly points out, these two terms refer to entirely different phenomena. Evolution is a historical process of change, just like the fall of Rome, and Darwinism is one of many theories—Lamarckian, theistic, emergentist, progressive creationist, and so forth—explaining how it came about.

An analogy can be drawn with the various theories—moral decline, the pressure of the barbarian invaders, the rise of Christianity—that have been proposed over the centuries to explain the fall of classical civiliza-

---

[1] Phillip E. Johnson, *Defeating Darwinism by Opening Minds* (1997); *Reason in the Balance* (1995); *Darwin on Trial* (1991)—all published by InterVarsity Press, Downers Grove, Ill.

[2] *Christianity Today* (8 December 1997).

[3] See the opening chapters of this book.

tion. As paleontologists have documented the process of biological change over time, historians like Gibbon have documented the fall of Rome year by year. And just as historians from Gibbon down have argued endlessly about the reasons for Rome's fall, so biologists have argued just as endlessly about the cause of evolution.

Unfortunately, the two terms are often used interchangeably and this causes confusion. The confusion is serious because it creates the impression that there is a considerable body of biologists sceptical of evolution when they are only sceptical of a particular evolutionary mechanism—Darwinism.

Johnson is certainly guilty of misusing the terms, but then, so are many biologists including, I must confess, myself for having entitled a previous book of mine *Evolution: A Theory in Crisis* when a more appropriate title would have been Darwinism: A Theory in Crisis.[4] Moreover, in sections of the book I often use the term evolution or evolutionary model when I should have used Darwinian or Darwinian model of evolution. The book was intended to be an attack on the Darwinian claim that all evolution can be plausibly explained by the accumulation of successive small random mutations. In the last paragraph of the book I summarized its essential theme in the concluding statement that "nature refuses to be imprisoned" within the confines of Darwinian thought. However the book was not intended to support special creationism. In the last paragraphs of the first chapter "Genesis rejected" I wrote that, "the world bore no trace of the supernatural drama that Genesis implied," that the special creationist framework "was frankly non-scientific and irreconcilable with the fundamental aim of science to reduce all phenomena to purely natural explanations."

If I had always used the terms Darwinism and evolution more carefully, much confusion could have been avoided. Nonetheless, throughout most of the text the terms are used correctly and there should have been no doubt that the book was intended primarily as critique of classic gradualist Darwinism. Ironically both creationists and Darwinists, for their own different reasons, often found it convenient to read 'evolution' for 'Darwinism' in *Evolution: A Theory in Crisis*—the creationists to find support for their antievolutionism and the Darwinists to claim the book was antievolution rather than anti-Darwin.

---

[4] Michael Denton, *Evolution: A Theory in Crisis* (Bethesda: Alder and Alder, 1986).

Michael J. Denton

The Significance of the Gaps

To a very large extent the arguments of Johnson, and indeed of special creationism throughout the past 150 years, depend critically on the claim that the gaps between different groups of organisms are absolute, could not have been be closed via a series of functional intermediates, and are *prima facie* evidence against common descent and can be taken as evidence for divine intervention.

A primary problem with this stratagem is obviously, How can we be *absolutely sure* that the gaps are as real as they may appear? If there is *even the slightest room for doubt*, the whole stratagem collapses. And one reason for doubt is, as Lamoureux points out, that gaps that once seemed unbridgeable have been closed as knowledge has advanced. The fact that some gaps, once considered irreducible, have now been reduced is damning, because it introduces an element of *irreducible* doubt into the debate regarding the reality of the gaps and their supposedly irreducible nature and effectively undermines the whole basis of the argument.

But there is however another far more serious empirical problem with this reliance on gaps that I want to try to discuss briefly here.

I think Johnson is right in arguing that Darwinism demands extreme gradualism. Darwin was himself insistent on gradualism, as witness the famous line from the *Origin "natura non facit saltum"*,[5] and so was Fisher and more recently Richard Dawkins.

Having concluded that Darwinism demands gradualism, Johnson then alludes to the gaps in the fossil record and argues that these invalidate Darwinism. If there are indeed significant gaps between different groups of organisms, and if indeed complex adaptations are not led up to gradually, then I think this is self-evidently a threat to classic Darwinism. I am sure that Richard Dawkins would agree. But unfortunately, and I think this is a key point, disproving Darwinism is not the same as disproving the theory of common descent.

Consider for example the recent advances in developmental biology that suggest strongly that the origin of the higher metazoan phyla occurred *per saltum* by the sudden redeployment of regulatory genes such as the Hox genes, or by sudden shifts in developmental mechanisms.[6] These developments certainly pose a severe challenge to the classical gradualistic Darwinism picture. In the case of the origin of the chordates, for example, there are some developmental grounds for believing that

---

[5] Nature does not make leaps (own translation)
[6] See *Nature* 389: 718.

the origin of the dorsal nerve cord and ventral aorta (the opposite of what is found in most invertebrate phyla) may have occurred in one sudden mutational step. This is a very far cry from the traditional idea that the major evolutionary transformations occurred ever so slowly via a long set of intermediate forms. Indeed, if this new picture is close to the truth then there may never have been any intermediates between some of the major phyla.

Moreover these developments are now raising other radical possibilities. In the case of the echinoderms, for example, it seems that even the main divisions within the phylum (equivalent to the mammals and birds in the vertebrates) may also have occurred suddenly via the redeployment of developmental genes.[7] In my view it is increasingly likely that the divisions between the main classes within the Mollusca and the Arthropoda and even, in the case of the insects, perhaps between different orders were also generated by the differential deployment of various development processes and genes. And even within the vertebrates it is possible — given the long-standing difficulty of imaging just how the aortic arches, for example, were transformed during vertebrate phylogeny via a series of tiny step-by-step changes — that these changes and much of the evolution of vertebrate soft anatomy (not recorded in the fossil record) were also brought about by major developmental shifts, *per saltum*. Lamoureux touches on the same point in his discussion of developmental genes such as the Hox genes and Sonic hedgehog.[8] These developments are arguably the most important in evolutionary biology this century. In fact it is interesting to note in passing that the emerging picture is strikingly reminiscent of the typological worldview of some of the leading pre-Darwinian biologists such as Richard Owen, who believed that the hierarchical pattern of the diversity of life was generated via an in-built program that involved the original animal archetype being successively subdivided, first into the major phyla, which were then successively subdivided into the lower taxa, and so forth.

These recent advances in developmental biology provide a very graphic example of why the existence of gaps, even if seemingly real and unable to be closed by conceivable functional intermediates, does not provide either evidence against common descent or justification for proposing divine interventions as the creationist's claim. *Gaps in short can no longer be taken as evidence against common descent.* Indeed just the reverse.

---

[7] Ibid.
[8] See page 19 of this book.

For ironically, because it seems increasingly that the gaps will eventually be explained in terms of changes in basic embryological processes and differential gene expression that seem increasingly analogous to processes that occur during ordinary development, the gaps between the major phyla such as the vertebrates and invertebrates may provide what is perhaps the *best evidence* to date for descent with modification.

These advances in developmental biology, together with the new knowledge of molecular genetics, imply that organisms that may look very different at the morphological level can be very close in terms of their basic genetic design, that is, at the level of their DNA sequences. In other words, the existence of a functional gap in phenotypic or morphological space does not imply an equivalent gap in genotypic or DNA space. Because all organisms and every aspect of their biology from the molecular to the organismic level are specified in a program embedded in their DNA, we have now, as I point out in my new book *Nature's Destiny*, in effect two quite different representations of a living organism — the gross familiar phenotypic representation and the invisible abstract DNA representation. Surprisingly many species and adaptations that seem to be very far apart in morphological space may be amazingly close in DNA space. For example, despite the extraordinary morphological differences between the species of cichlid fish in Lake Victoria in Africa, the differences in their DNA sequences are minimal — equivalent to the differences in the DNA between, say, an Australian Aborigine and a European. The same is of course true of the very dramatic differences between different breeds of dog — which are almost indistinguishable in their DNA. Again, the DNA sequences of man and chimpanzee are extremely similar, while the two species differ significantly at the phenotypic level in a variety of ways. The consequences of the existence of these two very different representations of the organic world can be illustrated by the following analogy. A contour map is a representation of a 3D topology on a plane surface. While it may be very difficult to move from one peak to another in real 3D space, it may be quite easy to perform the same move on the 2D map. An irreducible gap in phenotypic space cannot be taken to imply there is a similar gap in genotypic space.

Geographical Distribution

The credibility of a worldview or theory can never be wholly sustained merely because there are certain problems or deficiencies in an alternative competitor theory. An hypothesis is accepted over its competitor because it can explain most of the facts more parsimoniously and convincingly. When it comes to accounting for the facts of geographical distribution, special creationism is certainly far less convincing than its evolutionary competitor.

As Darwin comments in *On the Origin of Species*, one of the most remarkable facts about the inhabitants of oceanic archipelagos such as the Galapagos and the Canary Islands "is their affinity to those on the nearest mainland, without being the same species." [9] And as he rightly goes on to conclude, this pattern — which is found on every single island group throughout the world in varying degrees — is "utterly inexplicable on the ordinary view of the independent creation of each species, *but is explicable* on the view of colonisation from the nearest and readiest source" (italics my own). Here we have one of the classic examples of the way in which scientific theories are judged within the scientific community — namely on their relative ability to explain the facts.

While travelling in the Malay archipelago between 1854 and 1862, Alfred Russell Wallace (co-propounder with Darwin of the theory of natural selection) was brought face to face with those same facts of geographical distribution so suggestive of descent with modification that Darwin had earlier observed while on the Beagle. Four years before the publication of the *Origin of Species*, in February 1855, Wallace wrote a paper at Sarawak on the island of Borneo entitled "On The Law Which Has Regulated the Introduction of New Species." Wallace opens his paper with the sentence: "Every naturalist who has directed his attention to the subject of the geographical distribution of animals and plants, must have been interested in the singular facts that it presents." And he goes on to ask:

> Why are the genera of Palms and of Orchids in almost every case confined to one hemisphere? Why are the closely allied species of brown-backed Trogons all found in the East, and the green-backed in the West? Why are the Macaws and the Cockatoos similarly restricted? Insects furnish a countless number of analogous examples: the Goliathi of Africa, the Ornithopterae of the Indian Islands, the Heliconidae of South America, the Danaidae of the East, and in all, the most closely allied species found in geographical proximity.

Wallace also alludes to the fauna of the Galapagos:

---

[9] Charles R. Darwin, *On the Origin of Species*. Facsimile of the first (1859) edition, introduced by Ernst Mayr (Cambridge, Mass.: Harvard University Press, 1964), 388–406.

Such . . . phenomena as are exhibited by the Galapagos Islands, which contain little groups of plants and animals peculiar to themselves, but most nearly allied to those of South America, have not hither-to received any, even a conjectural explanation.

Wallace alludes also to the suggestive fact that:

When a range of mountains has attained a great elevation, and has remained so during a long geological period, the species of the two sides at and near their bases will often be very different, representative species of some genera occurring, and even whole genera being peculiar to one side only, as is remarkably seen in the case of the Andes and rocky Mountains.

Then Wallace makes another very telling point:

In all those cases in which an island has been separated from a continent, or raised by volcanic or coralline action from the sea, in a recent geological epoch, *the phenomena of peculiar groups or species will not exist.* (italics my own)

In other words, on young islands there had been insufficient time for evolution to occur.

From the facts of geographical distribution Wallace concludes his paper with the famous statement that "*Every species has come into existence coincident both in time and space with a pre-existing closely allied species.*" And he asks: "In the face of these facts the question forces itself upon every thinking mind—why are these things so?" And the answer that was in turn forced on both Darwin and Wallace was the concept of descent with modification, that is, organic evolution.

Finally consider and contrast the relative ease of the special creationist model and the evolutionary model to account for another fascinating case of geographical distribution, that of the living ratites (flightless birds) and the other unique species of animals and plants found only on the southern continents. The curious distribution of the southern flora and fauna was first noted by Banks on his voyage to the Pacific with Cook aboard *Endeavour* in the eighteenth century. How did so many obviously related types come to inhabit the widely separated southern lands?

I introduce the distribution of species on the southern lands here because the final solution to the mystery only came in the twentieth century with the development of plate tectonics, the evidence for continental drift, and the discovery of the great southern supercontinent, Gondwanaland. The solution represents a wonderful example where a prediction based on the presumption of evolution—that land bridges must have existed between the southern lands—was dramatically confirmed many decades later by advances in branches of knowledge unimaginable in the nineteenth century.

At present there are flightless birds in South America (rhea), Africa (ostrich), Australia (emu and cassowary), and New Zealand (kiwi). These species differ considerably in terms of their anatomy, behaviour, etc. but are clearly strikingly similar in many basic biological characteristics. If we are to explain this distribution in creationist terms, we must assume that God created each ratite species separately in each of the four regions in which they are found today. There is nothing exceptional in such a postulate of course, until we realise a very striking fact—that these four now separate geographical regions were once united in the great southern supercontinent of Gondwanaland more than 100 million years ago. In other words, the present distribution of the ratites corresponds to an ancient continental pattern now long vanished. How does the creationist and evolutionary explanation of these facts compare? To explain this pattern in creationist terms, we must assume that God has for some reason created each different ratite species on each of the now separate regions that were once part of Gondwana. To explain it in terms of common descent, we must presume that the several living species are all descended from a common ancestral flightless species that once inhabited the great southern land before it began to split apart. Now if it were only the distribution of the ratites that exhibited this pattern, we might perhaps find deciding on which explanation was the most plausible somewhat difficult. But it turns out that it is not only the distribution of the ratites which is 'Gondwanan' but, as Banks noticed in the eighteenth century, the distribution of many plants and other species on the southern continents also conforms to this Gondwanan pattern. So on the creationist model, we now have to assume that for some reason God must have created a vast biota of diverse but related forms and placed them on just those land masses that were once part of Gondwanaland. But why should God's creative activities in the present be constrained by an ancient and long vanished pattern of continental land masses?

The relative implausibility of the creationist model grows further when we examine the DNA sequences of the modern descendants of the ancient fauna and flora of the supercontinent. What we find is fantastically difficult to account for on creationist terms. By comparing the DNA of the various related species stranded in Australia, South America, and Africa as Gondwanaland fragmented, and extrapolating backwards using molecular clock estimates to the time when the sequences converge into ancestral sequences, we get a date of approximately 100 million years. This is much the same date that we derive from geological and geophysical evidence for the initial splitting of the supercontinent. If the creationist explanation of the spatial diversity of the species of the southern lands is accepted (i.e., that God for some reason created a vast class of related animals and plants only on those lands that had millions of years previously been part of Gondwanaland), we must now also accept that the differences in the DNA sequences of the southern flora and fauna were also specifically contrived so that they exactly correspond, according to the molecular clock calculations, with the time when the supercontinent split up.

I think the relative plausibility of the creationist and evolutionary accounts of the spatial and temporal pattern of the unique flora and fauna of the southern lands speaks for itself.

I make no apology for having laboured to some degree this discussion of the facts of geographical distribution. Creationists often claim that the facts do not support the concept of organic evolution. However I believe it is incontestable that the facts of geographical distribution—as Darwin and Wallace saw so clearly when they first encountered these facts in the nineteenth century—are far easier to explain by evolution than by special creation.

I think in the face of *the facts* of geographical distribution, the inference to descent with modification is inescapable, and I suggest that if indeed special creation is true, then it is evident that God must have created life to appear as if evolution had occurred. But what sort of God behaves in a manner so as to deliberately mislead mankind as to truths about the natural world?

Human Biology

In passing it is worth noting that there are aspects of human biology that make a great deal of sense if modern humans are descended from a non-human primate ancestor but that are very difficult to account for by the special creation model. These phenomena include the path of the recurrent laryngeal nerve, the large size of the head of the human baby relative to the birth canal and the consequent difficulties of the human female during delivery, the grasp reflex of the human infant, and the possibility of choking because both air and food share the same passage (the pharynx).

In addition to our anatomy that echoes our primate origin, it is also worth noting that sequential comparisons of human DNA (applying the molecular clock) suggest that the last common ancestors of modern humans occurred about 500,000 ago, and that the last common ancestors of humans and chimpanzees occurred about 5 million years ago. These genetic estimates correspond reasonably with the traditional dating of these key events based on the study of fossil evidence.

Design and Special Creationism

Despite my misgivings about the creationist ideology I certainly agree with Johnson, as does Lamoureux, that the cosmos exhibits evidence for design. However I do not believe that adaptive design exhibited by living organisms necessitates special creation. Water exhibits innumerable adaptations for carbon-based life, but I do not see that this implies water to be specially created. Similarly the earth exhibits a stunning variety of adaptations that are supremely fit for a complex biosphere like our own, but these adaptations are not the result of special creation but rather the outcome of millions of years of planetary evolution. Interestingly Paley's classic *Evidence*[10] presents a vast number of instances in nature where objects exhibit means-to-end complexity, yet Paley hardly ever touches on the actual historical process by which such design was executed. I am not saying, of course, that Paley was an evolutionist, only pointing out that his argument does not depend critically on how the designs he mentions were historically realised.

Of course not everybody accepts that the 'adaptive design' of living organisms is the result of intelligent activity. There has always been an alternative atheistic explanation for design: the appeal to an immense or infinite time during which chance plus selection would have had ample opportunity to generate objects that give the appearance of intelligent design. This explanation for adaptive design was first proposed by the ancient Ionian philosophers—the founders of the western intellectual tradition—and has been used by atheists ever since. According to the ancient atomists in an infinite universe, or in a universe that is infinitely old, or where there are an infinity number of universes, or where one universe goes through an infinite number of cycles, every single kind of material assemblage and every conceivable event, however improbable, is bound to occur and occur over and over again an infinite number of

---

[10] William Paley, *A View of the Evidence of Christianity* (1794).

times. This was in essence David Hume's escape, it was Darwin's escape, and also the escape of many modern cosmologists. I do not find this argument aesthetically appealing or convincing. Nonetheless I do think that once infinity is granted, then chance and selection could in theory generate any object allowable by physics—even one that gave the overwhelming appearance of design and appeared by every judgment 'vastly improbable'.

Philosophical Supernaturalism

Johnson is opposed to and vigorously attacks what he calls philosophical naturalism, but he seldom stresses that his own worldview is its logical antithesis—philosophical supernaturalism—a worldview that, if taken seriously, would render impossible any coherent understanding of nature.

It is true that, ever since the scientific revolution in the sixteenth century when Aristotelian and teleological explanations were rejected as sterile guides to genuine scientific explanation, naturalism as it is usually applied in modern science has been neutral with regard to the ultimate purposefulness or otherwise of nature's order and the laws that regulate it. Nevertheless, there is no fundamental antagonism between naturalism and a belief in a purposeful cosmos in which God's ends are achieved by the operation of natural law. Science since 1600 has been concerned with 'how things work' rather than 'why things are the way they are' because this focus leads to advances in knowledge while the 'why' approach is generally more sterile. But this does not mean that science is fundamentally opposed to teleological speculation. Indeed many of the most significant contributions to scientific knowledge were made by scientists such as Newton, Boyle, Faraday, and Clark Maxwell who were convinced theists or believing Christians.

I would concede that operationally the great majority of all working scientists are day-to-day philosophical naturalists, that is, they discount the occurrence of supernatural explanations—as indeed does the general public in most matters of mundane concern such as seeing a doctor, designing a plane, or planning a holiday. However, I think that Johnson exaggerates the supposed influence of philosophical naturalism in science. On the whole scientists are far more interested in the mundane business of getting results than holding or defending particular philosophical positions. The fact that so many working scientists believe in God indicates, as Lamoureux points out, that many keep much of their philosophical naturalism—if it is indeed the challenge that Johnson claims—safely locked in the lab.

My own philosophy might be described loosely as close to philosophical naturalism. I describe it unambiguously in the Note to the Reader in the beginning of my new book *Nature's Destiny* where I state my belief

> that the cosmos is a seamless unity that can be comprehended in its entirety by human reason and in which all phenomena, including life and evolution and the origin of man, are ultimately explicable in terms of natural processes.

I am not a philosopher and not competent to split philosophical hairs, but I would have thought this is close to what Johnson terms philosophical naturalism. But as will be evident to anyone who reads the book such a philosophy is not incompatible with a teleological interpretation of nature, nor does it lead necessarily toward an atheistic worldview.

As is evident from *Nature's Destiny,* I view life and man as an integral part of nature. I reject completely the special creationist worldview that organisms are in essence artifact-like and that God assembled different living things as an engineer might assemble human artifacts. On the contrary I see the entire course of evolution as *driven entirely by natural processes and by natural law.* I view the entire natural order as a designed whole with mankind as its end and purpose and with all the laws of nature specifically contrived to that end. Rather than being an artifact, something apart from nature, I view mankind as the very essence of nature and the natural end of the cosmic order. Turning to more specific points, I am also inclined (as I was when I wrote *Evolution: a Theory in Crisis*) toward the typological view that most biological forms are built into nature or 'robust' as Stuart Kauffman and Brian Goodwin term them. I also think, as already alluded to above, that the course of evolutionary change was in many instances saltational rather than gradual, and that many different groups of organisms were generated by changes in the deployment of developmental genes and processes and not via the gradual accumulation of a succession of small changes. I remain entirely opposed to the notion that a succession of random undirected changes in biological systems was responsible for the evolution of life.

The special creationist worldview, being in essence supernaturalistic, denies itself the possibility of any genuine teleological interpretation of man's centrality in the natural order. By dividing the world into two different realms—the natural and the supernatural—both of which are essential to bring about the ends of God – special creationism renders impossible any rational allusion to the laws of nature as specifically purposed to bring forth man. According to special creationism, man needs both nature and supernature so no strictly rational and entirely natural demonstration of man's centrality in nature can ever be achieved.

The adoption of a special creationist, supernaturalistic creed ironically precludes for ever the development of a fully rational demonstration of man's centrality in nature, or indeed any argument from the natural order purporting to demonstrate purpose in the cosmos. If end X is the product of natural laws Y, plus supernatural events Z, then X can never be inferred to be the purpose of Y. The argument can never be secured. Only if Y is entirely sufficient to explain both the being and becoming of X, can X be inferred to be an end of Y.

And these considerations lead to another serious problem with supernaturalistic explanations. Even if we could prove that an adaptation was so improbable that it must be the result of intelligent design, the most we can argue for is an intelligent designer. But we cannot infer that the intelligence behind the design is the creator of the universe – God. There are other candidates; living things might themselves be intelligent or, as James Lovelock has proposed the biosphere as a whole might be an intelligent being – Gaia, the Earth Goddess. Another candidate might be an extraterrestrial race who could conceivably have been involved in terraforming the earth over the past several billion years. Such activity would require intervention in the course of nature on earth and may have resulted in all sorts of complex adaptations.

From the above it is clear that purpose or design in nature that is the result of natural law can be safely attributed to the author of those laws – God. All purposes and designs in the cosmos that are not the result of natural law can never be safely attributed to God.

Conclusion

In conclusion, I agree with Johnson that the Darwinian model is an inadequate explanation for how evolution occurred. And I think he is right to attack the exaggerated claims of certain Darwinian theorists who extend Darwinian explanations to include all aspects of human nature and behaviour. Where he does this I applaud his efforts. I also agree with him that the living organisms exhibit design. However I am not aware of any convincing arguments put forward by Johnson to show that this design necessitates special creation. I am also unaware of any serious systematic attempt by Johnson to show how the facts of biology, such as those of geographical distribution discussed above, can be accounted for *more* plausibly in creationist than evolutionary terms. Until he does this, academic biology will not take his antievolutionism seriously.

In his advocacy of special creationism I believe Johnson is merely the latest in a succession of vigorous creationist advocates who have been very influential within conservative Christian circles, particularly in the United States, during the twentieth century. None of these advocates, however, has had any lasting influence among academic biologists. This is not because science is biased in favour of philosophical naturalism but because the special creationist model is not supported by the facts and is incapable of providing a more plausible explanation for the pattern of life's diversity in time and space than its evolutionary competitor. The reason why no current member of the US National Academy of Science is a special creationist is because of the facts, the same facts that in the nineteenth century convinced Darwin and Wallace and all the leading Christian biologists, including Joseph Hooker, Asa Gray, and Charles Lyell, of the reality of descent with modification.

Of Apples and Star Trek, Guidance and Gaps

Rikki E. Watts

I have had the pleasure of being hosted on separate occasions by Phillip Johnson and Denis Lamoureux. I found both charming and genial, and count them among my friends. It is therefore more than a little difficult to be asked to comment on their present disagreement: one has the awful feeling of being caught between shifting tectonic plates. But nevertheless I am more than happy to entrust myself to their grace and understanding.

Denis Lamoureux in a wide-ranging response has taken issue with Phillip Johnson at a number of points. However, it seems to me that the substantive question is quite straightforward. Both sides affirm design, but they differ as to how it is effected. Lamoureux argues that design is introduced through unobservably guided natural processes which then give rise to the complexity of life, a process that Howard Van Till regards as being due to the earth's being optimally gifted. Johnson argues that the complexity of living things cannot be accounted for by any known natural process.

I have difficulty understanding what Lamoureux means by unobservably guided natural processes. For example, how do guided natural processes differ from unguided natural processes? I am not sure that Lamoureux fully explains this. The water plunging from a waterfall into a canyon and the water fired from a water cannon into a crowd of protesters will produce 'natural' effects consistent with the qualities of water, gravity, etc. But I doubt that the protestors would be under any illusion as to which set of effects were due to guidance. The protestors can see the arrival of the cannon — itself designed — at the site, its being maneuvered into position, and its being directed at them and not the riot police. But this is not what Lamoureux has in mind. In his unobservably guided natural processes no such signs are present, not even the cannon itself. Instead we should probably consider Newton's famous apple. It falls to the ground under the influence of gravity and strikes Newton on the head. Lamoureux, if I understand him correctly, could regard this as an example of unobservable divine 'guidance' (of the apple, not Newton) through natural forces.

If this is so, it seems to me that Lamoureux's description involves, if not a semantic sleight of hand, at least a semantic confusion. First, I should have thought that the whole purpose of using the terms 'guidance' and 'natural processes', in the context of this debate, was in fact to distinguish between them. Throwing them together sounds very much like trying to have your cake and eat it too. Second, inherent in the idea of guidance is the notion that a particular event or series of events has a very low probability when considered within the context of the highly probable outcomes of natural processes. How does one make sense of a statement that claims an event or series of events has, at one and the same time, a relatively high (natural processes) and a very low (guidance) probability? How does this differ from speaking of square circles?

Related to the idea of guidance is the notion of intelligent intervention. It is crucial, however, that one avoid a deist conception of reality whereby autonomous natural processes are set against divine, 'miraculous' intervention. The heart of the issue is autonomy. Although the language of natural laws is helpful in that it suggests order and regularity, it becomes problematic if it implies some kind of inherently absolute status whether rooted in the natural world itself, or, to use Spinozan terms, in divinely necessary decrees. On these models, any intervention in or manipulation of natural processes is either an offence against some absolute or an impossibility since it requires that God transgress his own decrees.[1]

But there are several problems here. First, the relationship between intervention and miracles needs to be explored. My first degree was in aeronautical engineering and I can assure readers that 747s are not the product of natural processes, time, and chance. They clearly involve intelligent manipulation of and intervention in natural processes. But I doubt very much if my fellow engineers would define the process as miraculous. On the other hand, there is nothing inherently mysterious about my contact with the world. It is doubtful if the same could be said of God's involvement, not least since he is spirit.

---

[1] Spinoza, recognizing the mistake in seeing nature as autonomous and God as inactive except when he miraculously intervened, argued against miracles on the grounds that since God's decrees constantly undergird nature, miracles would mean that God contravened his own divinely necessary decrees. His whole point rests on the rhetorical move of granting nature a Calvinist decretal status (using the then-pervasive Calvinist language of the council of Dort). See Colin Brown, *Miracles and the Critical Mind* (Grand Rapids: Eerdmans, 1984), 30-32.

Second, the Hebrew Bible knows nothing of 'nature' in the autonomous sense of the Greek *physis* (the first occurrence is in 3 Maccabees 3:29). In the ancient Hebrews' mindset, the coherence of the present order is the immediate consequence of the sustaining Word of the Great King who, in faithfulness and compassion, maintains the good order of his temple-palace cosmos for the sake of his children (cf. Isa. 66:1). Neither autonomous status nor divine necessity is granted to nature. So-called 'interventions' can hardly constitute a breach of autonomous or necessary natural laws. Instead, they grow out of precisely the same fundamental compassion and faithful love of God for his creatures which undergirds the predictability and constancy of the natural world. The only difference is that while some actions by their very regularity are quickly taken for granted and almost lapse into the cliché of 'natural processes', more suprising acts of God by their very irregularity and striking nature are easily regarded as interventions. But their striking nature is in part their point. They are intended to impress themselves on human consciousness. This is why the biblical writers describe such moments either as signs to instruct or encourage (*'ôt/sēmeion*) or as powerful acts to deliver God's people (*gᵉbûrâ/dynasteia/ischys/dynamis*), often performed in wondrous and awesome ways (*môpēt/terata*). This is also why, while on the one hand 'interventions' are seen as awesome 'deviations' from the common experience of the natural order, they can also be prayed for in that they reflect Israel's past experience of God's steadfast love. In other words, what we are speaking of, at bottom, should probably be seen not so much as a difference in kind, but a difference in degree. Thus far I think Lamoureux would agree. It is ironic, however, that the compassionate faithfulness and dependability of God, which underwrote Western science, is exactly the thing which, when it is most evident (in intervention), is excluded by that very science to which it gave birth.

But this is not all. Interestingly, these signs and powerful acts tend to cluster around a few major moments, the creation and the exodus, with the latter also being construed as a new creation. Allowing this biblical analogy (exodus as creation), if the exodus also entailed various moments of intervention then this might raise some questions for the deist who, in principle, allows only one intervention and then insists that God desist from then on. I am happy, then, to employ the language of natural processes, guidance, and at a pinch, intervention, provided we understand what it means. This also avoids the idea that God is sometimes present and sometimes absent from his creation. In my capacity as a professor I preside over seminars. I may or may not directly intervene, but my presence is still there in a sustaining capacity. Likewise, God is always present in his role of sustaining what we might call natural processes, but he may also occasionally choose to intervene more directly in the sense of guidance. On this view, when we examine Lamoureux's statement closely it appears that he is confusing God's providential ordering and undergirding of his creation (evident in what we call natural laws) with the more personal intervention implied by the word 'guidance'. If so, we may ask in what sense is it meaningful to speak of guidance or design when one allows only the operation of natural processes?

This raises another question. What exactly does Lamoureux mean by a guidance that is inherently unobservable? At the very least he seems to think that it should result in no observable dislocation of natural processes. It is important to note here that he does not believe that if only we had better and more advanced means of observation then we could detect this guidance. Not at all. He means that this divine guidance, by its very nature, is forever beyond scientific observation. In other words, the attribution of 'guidance' to physical processes is predicated solely on faith. It is here that he invokes a surprising analogy with the blood of Jesus. He argues that although we might potentially examine the chemical make-up of Jesus' blood, scientific observation would never reveal its efficacy in dealing with sin. For Lamoureux, just as the efficacy of Jesus' blood is accessible only to the eyes of faith, so too God's guidance of natural processes is, in principle, inaccessible to scientific verification. They are two different spheres, and they are quite separate.

But this creates (no pun intended) two further problems. First, the analogy strikes me as a poor one. It confuses *independent* significance *freely granted* with a significance *intrinsic* to the nature of something. Jesus' blood (to use Lamoureux's expression) is efficacious not because there is something inherently cleansing in it, but because God grants it a significance quite independent of any chemical composition. Given this, it is hardly surprising that this kind of quality is undetectable by scientific observation. In the nature of the case, no such significations ever are. On the other hand, the question of whether something was guided or not seems intrinsic to the nature of actions or events. Lamoureux, at least to me, again appears to be confusing categories.

This leads to the second problem. Christian theism has characteristically proclaimed its faith on the basis of observable historical events.[2] In

---

[2] Granting that the exact content of these events is inextricably intertwined with interpretation does not negate their intrinsic historical reality.

what sense can faith in a guidance which *by definition* can have no observable point of contact with the natural world call itself Christian? Here, too, Lamoureux's analogy seems flawed. The Christian affirmation concerning Jesus' death does not, in fact, stand utterly divorced from the realm of observation. In 1 Corinthians 15:17 Paul links our faith that we are justified by Jesus' efficacious death with the resurrection: "If Christ has not been raised, your faith is futile and you are still in your sins" (NRSV; cf. Rom. 4:25). The saving power of Jesus' blood (i.e., his death), while not an immediately observable quality, is nevertheless not discontinuous with the physical world (cf. 1 Cor. 15:1ff.). Its truth claim is related to an observable historical event, namely, the resurrection. Granted, the resurrected Jesus was not seen by everyone, but that is beside the point. It was clearly not a matter "for the eyes of faith only", as Thomas's case indicates (John 20:24-29; cf. Matt. 28:17). More problematic is that Lamoureux's "unobservable guidance" appears to run counter to his citing of Psalm 19:1-4 ("The heavens declare . . .) and Romans 1:19-20:

> [19]Since what may be known about God is plain to them, because God has made it plain to them. [20] For since the creation of the world, God's invisible qualities—his eternal power and divine nature—have been clearly seen, being understood from what has been made so that men are without excuse.

The latter is clear: creation bears the marks of intelligent design and this is so regardless of whether one sees with the eyes of faith or not. But, surely, inherent in perceiving design is the observation of those places where guidance is deemed necessary. I am at a loss to understand how the presence of design could be evident to all and yet those points at which guidance intervened be inherently inaccessible always and forever except to the eyes of faith. My question to Lamoureux is straightforward: in what, *observable* respect does his theistic teleological evolution through guided natural processes differ from an atheistic purposeless evolution through unguided natural processes?

I think this is Phillip Johnson's point: it is hard to see how an unobservable guidance through natural processes differs from no guidance and only natural processes. Does not the former merely collapse into the latter? How does it escape the charge of being merely a semantic move without genuine content? Assertion does not constitute argument any more than abiding by grammatical rules ensures that a phrase makes sense, and appeals to Christian faith are hardly convincing if the faith called for does not appear to be all that Christian.

Perhaps, then, we are justified in restating Lamoureux's view as one that affirms the providential ordering and sustaining of natural processes such that they result in life. This is the second point at issue. Phillip Johnson claims (as does Michael Behe) that life's irreducible complexity is at present quite inexplicable on purely naturalistic terms. And we must be clear, Lamoureux provides no direct evidence that it might be otherwise. But it is not as if he avoids the issue. He responds in two ways. First, he seems to appeal to the fossil record and then offers a very brief comment on gene manipulation suggesting that it "could . . . conceivably result in dramatic morphological change". Leaving aside the difficulties inherent in shifting interpretations and classifications of the fossil record, both the record as it stands and language like "could . . . conceivably", even with the most generous interpretation, are still a very long way from demonstrating a mechanism. This is precisely Johnson's and Behe's criticism. Edit out the hand-waving and just-so stories, and it is immediately apparent that no one has even come close to showing how changes caused by natural processes alone "could . . . conceivably" result in the complexity of living things.

Although Lamoureux urges that God can and does intervene in the unpredictable and unique world of human activity (personal interventionism), he finds such interventions unlikely in the creation of the natural world (cosmological interventionism) since it is predictable, governed as it is by natural laws. I am unsure what to make of this distinction, not only because the vast majority of 'personal' biblical miracles are themselves 'violations' of natural and predictable laws, but also because I doubt if such a dichotomy can be sustained. Nevertheless, I think Lamoureux's concern is with the idea of a God who injects new information into the system, whether once or constantly (Lamoureux does not say which).

Although a minor point, 'constantly' can give a false impression. It is important to remember that we are speaking of many millions of years and it is probably wisest not to telescope them too much. But whatever the precise details, the more important question arises: how does Lamoureux know *in advance* how God must act with respect to the origins of complexity in life? How does he know with such certainty that God did not at various moments more directly intervene (especially given the biblical analogy of Exodus as creation)? It seems to me that at this stage in our knowledge we just do not know enough to be able to make statements like this. In other words, Lamoureux's assertion is based on a theological and philosophical *a priori* which amounts, in terms of the origins question, to a practical deistic naturalism. Now, there may be an aesthetic element to Lamoureux's *a priori*: it does seem a lot tidier if only natural processes apply. However, Lamoureux's distinction between personal and cosmological interventionism might well be his Achilles' heel. There is little question that the inorganic world is tidy. But life—all life and not just human life—is hardly so, and it may be this very unpredictability that rules out an appeal to aesthetic tidiness and so points to origins beyond the mundane regularity of the inanimate world. However, whether or not Lamoureux's assumption is correct, the important thing is that we recognize it for what it is: a philosophical and theological assumption. The same *a priori* undergirds Howard Van Till's "gapless and robust" formational economy theory in which he assumes that matter contains what is necessary to organize itself into the specified complexity necessary for life. How does Van Till actually know this? What direct concrete scientific evidence does he offer for this statement? None that I can see. How then can he too be so certain?

Apparently it all boils down to the God-of-the-gaps argument. Lamoureux argues that past appeals to divine intervention in the face of then-inexplicable phenomena have resulted in an embarrassing retreat when the gaps have been filled. This is indeed an important point. If past experience has shown that many things we thought demonstrated interventionism were in reality the result of natural processes, then caution is certainly in order before rushing to do so with regard to the origin and complexity of life. Yet I wonder if it is as simple as all that. As Van Till himself rightly recognizes, there is a significant difference between the problematics of self-organization of the physical world in general and those concerning complex biological forms in particular, even if he ultimately hopes that the strategies that worked with the former will prove fruitful with the latter. But how well founded is this hope?

First, it seems to me that the many gaps science has successfully closed reduce to one basic kind: descriptions—not explanations in the strict sense—of the low-level, self-organizational order inherent in physical and chemical interactions (included here are minor evolutionary variations). In one very important sense, then, it is misleading to imply a vast array of closed gaps when they are all only manifestations of the one gap, namely, our uncovering of the 'natural laws' underlying the regularity of the physical universe. This 'multiplicity' constitutes one gap and one gap only. But the ordered structure of the physical universe is of an entirely different (lesser both qualitatively and quantitatively) degree of complexity than the specific informational structures inherent in living things and their complex organs. That is, describing the mundane, systematic, and repetitive function of orderly natural processes is not at all the same thing as explaining the origins of complex (and quite unpredictable) information. They are, to use a colloquialism, not even in the same ballpark. The one has really nothing to do with other; indeed, if anything the two are inversely related.[3] Natural, orderly laws actually conspire against the emergence of complex information. For this reason I fail to see how many instances of closing the one gap have any bearing at all on the closing of the other gap of the origins of, and the specified informational complexity inherent in, living things. The two (not multiple) gaps concern entirely different classes of reality. To hope for success in one class because of success in another is, to my mind, a *non sequitur*.

Now it is important to understand exactly what this means. Those who use the God-of-the-gaps defence have apparently been so impressed both by the number of gaps closed and by the fact that describing natural processes has sufficed that they have assumed that all gaps can be so closed. That is, they have accepted an essentially naturalistic view of the way in which the universe originated. The problem is that being so impressed by sheer numbers they have failed to see that the many instances are really only of the one kind. And secondly they have not sufficiently grasped the considerable discontinuity between this quite mundane and rather basic, even repetitive, kind of natural interaction, and the hugely more complex issue of the origin of specifically complex information. An easy mistake to be sure, but a mistake nevertheless.

Van Till, seeking to undermine the notion of this second gap, argues that if such a gap exists then God must have chosen to withhold certain formational gifts from creation. If Johnson is right, he says, then God has equipped creation with an incomplete formational economy. This is a powerful argument since most Christians would recoil from the suggestion of God creating something that was incomplete. But Van Till's statement is problematic. First, it strikes me as tautologous since it assumes that irreducible complexity can arise by natural causes, the very point at issue. Second, and more importantly, what if it is inherently impossible for irreducible complexity or specific information to arise by means of uniform laws, chance, or a combination of both? Uniform laws result in uniformity (which is why science works as well as it does).

---

[3] See Polanyi and Dretske cited in Stephen C. Meyer, "The Explanatory Power of Design: DNA and the Origin of Information," in *Mere Creation*, ed. William A. Dembski (Downers Grove, Ill.: InterVarsity Press, 1998), 133.

Chance creates irregularity. But putting them both together does not lead to complex information. This seems to be Stephen Meyer's particularly potent contribution: as far as we presently understand, informational complexity of the language variety by its very nature simply could not, and therefore cannot, be generated by the operation of natural laws.[4] In other words, God could no more create complex-information-generating natural law than he could a four-sided triangle. The problem has nothing to do with God's power or goodness, and everything to do with the intrinsic nature of the stuff itself. If so, to describe this 'gap' in terms of an incomplete formational economy is simply to misunderstand the essence of irreducible complexity and specific information.

Intriguingly, in appealing to an Augustinian notion of an optimally gifted creation, Van Till also resorts in the end to a theological and philosophical *a priori* rather than science.[5] What we miss, as with Lamoureux, is a scientific discussion of the problem of how irreducibly complex structures or specifically complex information could arise by natural processes. With all respect, the irony is almost delicious. While our two scientists ultimately step outside their fields to invoke theology and faith, it takes a lawyer (also stepping outside his field) to ask: where is the science? It seems to me that Johnson has a point: what we have is a particular brand of philosophical and theological precommitment—a form of cosmological deism—telling science what it can and cannot propose. I do not mean to be provocative, but the question forms in my mind: have we not already been through this before?

Van Till also criticizes intelligent design theorists because, having addressed the conceptualization side of design, they have not yet proposed a viable model for how design is actualized. This is a valid concern. But two comments seem in order. First, it strikes me as unfair for Van Till to describe a one-hundred-year-old research program, supported by massive resources and thousands of the best brains, as a "theory in adolescence" and so to appeal for patience as it struggles to explain the origins of irreducible complexity, and then to dismiss a program less then ten years old, with comparatively minuscule resources, because it has not yet satisfactorily resolved what is undoubtedly a major issue. If anything, a reversed attitude has more warrant.

---

[4] *Ibid*, 132-4.

[5] For a critique of Howard's reading of Augustine at this point, see Jonathan Wells, "Abusing Theology: Howard van Till's Forgotten Doctrine of Creation's Functional Integrity," *Origins & Design*, 19 no. 1 (Summer 1998): 16-21.

Second, the question of describing the actualization of intelligent design and the failure of the natural processes model to account for irreducible complexity are, quite frankly, unrelated. That there are questions as to the best viable alternative energy source does not in any way lessen the fact that burning coal causes pollution. Whether or not the intelligent design movement can come up with a viable account of design actualization in no way diminishes the failure of the regnant natural processes model to explain irreducible complexity. And make no mistake. This is no peripheral matter, a minor blip soon to be resolved. It strikes at the very core of the whole enterprise. There might be a great deal of fuss in the box office, a forest of colourful hoardings with clever slogans advertising the greatest show on earth, and even much concentrated discussion as to the shifting of sets on the stage. But there is no script. An important question needs to be asked: what evidence would proponents of the natural processes model be willing to accept as demonstrating that their model as it presently stands is fatally flawed? If this cannot be answered clearly then I think we are justified in asking the unfortunately sharp but necessary question of whether we are dealing with genuinely free scientific inquiry or ideological strictures.

So how is intelligent design actualized? How is it to be related to the fossil record? Who knows. It is much too early to tell. Lamoureux and Van Till may find "punctuated naturalism," as Van Till characterizes intelligent design, unattractive. But I cannot see how preference comes into it. We all know of an exceptionally gifted scientist who once rejected quantum theory, not with a scientific argument, but with a retort along the lines of "God does not play dice with the universe." Unfortunately (or fortunately!) what we might prefer God to do does not always comport with what he actually does—a crucified Messiah and quantum physics being cases in point. It seems to me that the best science is always born of going fearlessly where the data leads, and of asking hard questions even when they challenge prevailing orthodoxy. It is possible that Johnson et al. have missed something. If so, at least we would know that that avenue did not work, and a closed door is also an advance. On the other hand, what if he is right and the origin of information is precisely that boundary condition (or event horizon) at which methodological naturalism breaks down? Whatever the case, surely the only way forward is to do our best with the facts before us and to be able to do so without the dead hand of some inquisitorial ideology dictating in advance what our findings must be. Perhaps *Star Trek* has got it right. What we need is a spirit of open inquiry that boldly goes even where most for whatever reasons—including fears about past 'closed gaps'—are hesitant to venture. I find my sympathies with Jean-Luc Picard who, having charted his course, begins the adventure by declaring: "Make it so!" Intriguing is it not, how similar that command sounds to one account of how this much larger adventure of which we are all part got going in the first place? But then, on reflection, perhaps it is not so surprising after all.

Does Methodological Naturalism Lead To Metaphysical Naturalism?

Loren Wilkinson

> There is nothing that God hath established in a constant course of nature, and which therefore is done every day, but would seem a Miracle, and exercise our imagination, if it were done but once; Nay, the ordinary things in Nature, would be greater miracles, than the extraordinary, which we admire most, if they were done but once. . . .though he glorifie himself sometimes, in doing a miracle, yet there is in every miracle, a silent chiding of the world, and a tacit reprehension of them, who require, or need miracles.
>
> —John Donne, *Easter Day Sermon*

The argument among Christians between proponents of 'intelligent design' and the (in principle) complete explanations of 'methodological naturalism' requires careful thought about both the character of science, and the character of God's relationship to creation.

Christians believe that the cosmos is the purposeful creation of a loving God. But, far from being incompatible with science, this belief about creation has encouraged the conviction that the cosmos is intelligible- and that it is appropriate for human beings to explore that intelligibility. These beliefs seem to have been the necessary (and culturally unique) precondition for science (science understood asin Kepler's words "thinking God's thoughts after him."). In any case, the deep compatibility of Christianity and science in its formative early centuries (the sixteenth to eighteenth) is almost beyond dispute.

From the earliest days of science, scientists who were Christians (Kepler, Galileo, Harvey, Maxwell, et al.) have seen no inconsistency in arriving at a completely naturalistic explanation of natural phenomena. (The apparent tautology—a *naturalistic* explanation of *natural* phenomena—is, unavoidable as long as we persist with the unbiblical category of "the natural") This was how God had created the universe to work. To describe its workings accurately was implicitly an act of worship and honour to God, though the description might make no mention of God.

But this originally reverent activity, in an increasingly secularized environment, had an ironic consequence. By the eighteenth century, to invoke the name of God in the explanation was implicitly to question his workmanship. Newton felt the only role left for God in his celestial mechanics was to intervene occasionally to set straight occasional irregularities in the movements of the planets. As these movements came to be explained through the gravitational attraction of other planets, this role for God was no longer necessary, though the system as a whole could still be seen as God's workmanship. When (In a famous exchange) Napoleon asked the astronomer LaPlace if there was room for God in his system, Laplace is said to have replied, "I have no need for that hypothesis." La Place's answer was not necessarily about the existence of God. For "I have no need of that hypothesis" is an answer consistent *both* with an affirmation of the excellence of God's creation, and a denial of belief in God. Perhaps as we shall see, the ambivalence is necessary to good science. To say simply "God did it" and to cease wondering will take us no further in our knowledge of the created world.

Nevertheless, the relationship of God to creation implicit in the faith of these early scientists became increasingly deistic: God is pictured as the great craftsman who left the elaborate clockwork of creation to function flawlessly on its own. It was in discovering the rules built into creation by that absent God that we were (in Kepler's phrase) "thinking God's thoughts after him." God was perceived to be totally transcendent; there was no immanence in such a picture. The idea of God's intervention in creation — in revelation, and in the miraculous — came increasingly to be viewed as an embarrassing mistake made by humans not willing to think. We could know all that we needed to know — about both Creation and Creator — through our own unaided reason and empirical observation.

Thus the deistic picture of a purely transcendent God, active at the beginning of creation but with a hands-off attitude ever since became popular during the eighteenth century. A part of the original deistic program of Lord Herbert of Cherbury was the denial of both revelation and the miraculous, and a confidence in reason's ability to uncover all that we needed to know about "the Supreme Being." It is no surprise that deism, with its distant and uninvolved deity, merged easily into atheism.

What is not so obvious is the unfortunate fact that deism — the belief in a designer God whose involvement with creation is only through a serious of flawless initiating acts, followed by 'natural' development — continues to be the framework in which much of the discussion of the relationship of Christianity and science takes place. In the intelligent design debate, which is the subject of this book, the principle difference seems to be whether God set all of the conditions right at the very first instant, or whether God intervenes occasionally to insert new design elements in the flow of 'natural' events. Both views, misrepresent the nature of God's relationship to creation — and hence, the nature of 'the natural.'

This picture of a God who is purely transcendent, a God who intervenes occasionally or not at all, is neither biblical nor Christian. Certainly it is in no sense Trinitarian, and leaves out the work of the Spirit (as in "You send forth your Spirit, and they [animals] are created" [Psalm 104:30]), and the Incarnate Word "without [whom] nothing was made that has been made" (John 1:31), and in whom "all things hold together" (Col. 1:17). Nevertheless, most Christians who have perceived a tension between religion and science have assumed a Creator-Creation relationship much closer to this unbiblical deist picture than to Trinitarian Christianity.

It is not surprising, therefore, (especially since Darwin) that there have been those who assume that a complete scientific description necessarily and completely excludes God from the picture.

Of course a complete scientific explanation of anything is extremely elusive. For something so complex as the origin and development of living things it is not remotely possible. In fact, such a description will always be unavailable — since, as in all research into past events, we are dealing with things that happened once, of which we have imperfect records (if any), and which are, by their very nature, unrepeatable. No meaningful laboratory experiment to test theories such as natural selection (on the macro level) or processes such as speciation can ever be done. Such experiments would require the duplication of a planet as a laboratory — and would take hundreds of millions of years for even one run-through. Whatever description we offer of such events will necessarily involve guesswork, extrapolation, and a good deal of faith, as does all science.

What is at issue, therefore, is not the *fact* of an elusive and ultimately unattainable scientific description, but rather whether the *ideal* of such a description is incompatible with the loving, personal, creator God revealed to us in Scripture and in Jesus Christ. Yet the ideal that complete understanding of a process necessarily excludes God from the picture contradicts our normal Christian practice. We regularly, for example, thank God for our food: rightly recognizing it as God's provision. Yet we could, if we took the effort, trace the corn or tomato back through many manmade and "natural" processes to its source. The practice of the "methodological atheism" of going regularly to the store (or the garden) to obtain such food does not necessarily produce "metaphysical atheism" in the eater, who still ought to thank God for his provision.

Nevertheless, in the area of biological origins, Christians and non-Christians still regularly assume an incompatibility between theistic belief and naturalistic description, but with different consequences.

'Incompatibilist' Christians have tended to avoid those scientific disciplines (biology, palaeontology, geology) whose practice assumes naturalistic assumptions about origins. Some of them have gone further and led an active campaign for 'creation science' to be taught in the classroom as an adequate description of the shaping of life, the planet, and the whole cosmos. Most members of this creation science movement assume a highly literal reading of the first chapter of Genesis, and are thus committed to finding evidence for a young earth. This usually leads to a form of catastrophism (rather than gradualism) which attempts to explain much terrestrial geology as being 'naturally' caused by the 'supernatural' flood described in Genesis. (However, not all those who are critical of the ideal of a naturalistic explanation are committed to catastrophism and a young earth. Phillip E. Johnson , for example in *Darwin on Trial*, has focussed on the gaps in the evidence for the Darwinian case, considered as a legal argument.)

Many incompatibilist scientists, on the other hand, have rejected belief in God, assuming that such belief necessarily implies the need to turn one's back on reality as it may be studied in the physical world. Many of these, overlooking the Christian origins of science, are simply reacting against what they perceive to be a necessarily anti-intellectual (and uncurious) strain in Christian orthodoxy.

Sometimes, however (perhaps often) their commitment to science is accompanied by an implicit rejection of the very possibility of God, particularly a personal God to whom they are ultimately responsible. Such practitioners of science have therefore a kind of faith-commitment to the non-existence of God, and are hopeful that a scientific description of origins will provide some sort of empirical validation for that commitment. They *hope* that methodological atheism implies metaphysical atheism. Incompatibilist scientists sometimes show as much zeal in their critiques of the notion of creation as incompatibilist Christians show in their critiques of evolution. In both cases, fundamental personal beliefs are being

challenged. Johnson's work has been helpful in uncovering these religious motivations of (apparently) non-religious science. His work is less
than helpful, however, when he seeks to show that the attempt to give a
completely natural explanation of origins is incompatible with belief in
an all-powerful Creator.

In contrast to these incompatibilists, on the other hand, many scientists who are Christians (and some who are not) have concluded that
there is no fundamental incompatibility between the ideal of a complete
scientific description and belief in God as revealed in Scripture and in
Christ. These scientists have a much more robust view of God's immanence; thus they have no difficulty in accepting the idea that the Creator
is at work in the whole process which can be described, from another
framework, 'naturalistically.' They tend to characterize a science (or a
faith) that must establish God's intervening, 'supernatural' action at a
particular point in the creation process (which thus would operate 'naturally' or 'on its own' the rest of the time) as depending on a "God of the
gaps."

C.A. Coulson, then professor of applied mathematics at Oxford, put
the matter succinctly many years ago when he wrote (in response to another Christian's description of the "deft touches" of God's occasional
interventions):

> . . . if God's action in nature is limited to "deft touches here and there" I
> can barely distinguish Him from the engineer who made the mecha-
> nism, and who leaves it to work its own passage, interfering only to put
> it right when something is going too far wrong. Either God is in the
> whole of Nature, with no gaps, or He's not there at all.[1]

As we have suggested, Christians and non-Christians who perceive
there to be a conflict between a naturalistic scientific description of origins, on the one hand, and belief in a creator God, on the other, are likely
to be operating from a sub-Christian, deistic picture that stresses God's
transcendence — his distance and otherness from creation — at the expense
of his immanence, his intimate involvement with each thing that happens.

And even those Christians who do not see such a conflict are usually
operating from an implicitly deistic framework, which is deeply en-
grained in the tradition of science — and perhaps in Western culture as a

---

[1] C.A. Coulson, *Science and Christian Belief* (London: Fontana Books, 1958), p. 35.

whole. There is perhaps a hint of such deism in Denis Lamoureux's otherwise cogent critique of Johnson, for he sometimes writes as though the scientist is investigating a process which has unfolded naturally from its initial divine origins. It would be better to make clear, as Scripture does, that the transcendent God is always immanent in the process (or perhaps, that the process is immanent in God– as in Paul's approving use of the Stoic phrase, "in him we live and move and have our being" [Acts 17:28], or the more direct, "in him all things hold together" [Col. 1:17]).

Those who have grown up thinking of God's relationship to creation in a deistic way often mistake any attempt to speak of God's immanence in creation as a kind of pantheism, a confusion of Creation and Creator. And it is true that some of those who try to develop a doctrine of creation appreciative of God's immanence do sometimes ignore his transcendence, and end up describing what is in effect pantheism. Such are many in the tradition of process theology. Sometimes they use the term *panentheism,* but in a way that suggests God and the universe are evolving together, with God emerging from the process (often in human consciousness).

The way to a healthy understanding of the relationship of God to creation and of Christian faith to science lies in the recovery of a more deeply Trinitarian Christian understanding of God, which affirms both God's transcendence and his immanence. No one has made this point more lucidly than Jürgen Moltmann, in *God In Creation*:

> The trinitarian concept of creation binds together God's transcendence and his immanence. The one-sided stress on God's transcendence in relation to the world led to deism, as with Newton. The one-sided stress on God's immanence in the world led to pantheism, as with Spinoza The trinitarian concept of creation integrates the elements of truth in monotheism and pantheism... God, having created the world, also dwells in it, and conversely the world which he has created exists in him. This is a concept which can really only be thought and described in trinitarian terms. [2]

The absence of such an understanding has haunted Western science and culture for a very long time and lies behind much of the creation—evolution debate—especially as it has been carried on among Christians.

---

[2] Jürgen Moltmann, *God In Creation* (San Francisco: Harper Books, 1986), p. 98.

Loren Wilkinson

Central to such an understanding is a recognition at the outset of the contingent nature of Creation. The upholding energies of the Creator are necessary *at every instant* for each thing to be. There is no question of God intervening in such a creation, because each thing depends for its very existence on God. If God were not in someway upholding the creature's existence, the creature would cease to exist. This knowledge of the contingency of creation is more vital in Eastern Orthodox Christianity than in the West. As one orthodox theologian, Philaret of Moscow puts it, "All creatures are balanced upon the creative word of God, as if upon a bridge of diamond, above them is the abyss of divine infinitude, below them that of their own nothingness."[3] Or, in the words of a contemporary Orthodox thinker, Kallistos Ware:

> As the fruit of God's free will and free love, the world is not necessary and not self-sufficient, but *contingent* and *dependant*. As created things we can never be just ourselves alone. God is the core of our being, or we cease to exist. At every moments we depend for our existence upon the loving will of God. Existence is always a *gift* from God.[4]

And again:

> As creator, then, God is always at the heart of each thing, maintaining it in being. On the level of scientific inquiry, we discern certain processes or sequences of cause and effect. On the level of spiritual vision, which does not contradict science but looks beyond it, we discern everywhere the creative energies of God, upholding all that is, forming the inmost essence of all things. But, while present everywhere in the world, God is not to be identified with the world. . . . God is *in* all things yet also *beyond and above* all things.[5]

If God is necessary to each thing's existence, language that speaks of God's intervening in a normally 'natural' process seriously misrepresents the nature of the Creator-Creation relationship. A Darwinian or evolutionary description is thus not necessarily a threat to our understanding of God's creative activity. A late-nineteenth century Christian defender of Darwin, Aubrey Moore, put the point succinctly when he observed that "a theory of occasional intervention implies as its correlative a theory of ordinary absence."[6]

---

[3] Metropolitan Philaret of Moscow, cited in Kallistos Ware, *The Orthodox Way*, (St. Vladimir's, 1979), 57.

[4] Ibid., 57.

[5] Ibid., 58.

[6] Aubrey L. Moore, *Science and Faith* (London: Keagan and Paul, Trench, 1889), 184.

The discussion about whether or not methodological naturalism leads to metaphysical naturalism is not relevant for the Trinitarian Christian, since strictly speaking there can be no "naturalism" at all for one who believes that all things depend upon God for their very existence.

Science, for Christians with such an understanding, like all knowledge, should be rooted in wonder, in the miracle that anything at all should exist. Miracle is exactly the right word. While scientific description describes casual relationships, nothing in the universe is ordinary or natural in the sense that it could exist without God's will and purpose.

This apparent paradox, of an action which can be described on one hand as 'naturally' and on the other as fully the work of God, is not unique to God's action in nature. It is evident, for example, in the process which Paul invites us into in it is one instance of the paradox that Paul describes in the second chapter of Philippians (v.12 – 13): "Work out your own salvation with fear and trembling, for it is God who works in you to will and to act according to his good purpose."

That verse describes one instance of a general pattern of God's relationship to creation: to work through the freedom of his creatures. That freedom is maintained at the very expense of the Creator; hence the centrality of the cross. As W. H. Vanstone put it, in *Love's Endeavour, Love's Expense*:

> We may say that Christ, the Incarnate Word, discloses to us at the climax of his life, what word it was that God spoke when "He commanded and they were created." It is no light or idle word but the Word of love, in which, for the sake of another, all is expended, all jeopardised, and all surrendered.[7]

Only when we look at creation (or 'evolution') from the standpoint of this costly immanence of the transcendent God will we begin to leave behind the deism that has long haunted our reflections on the relationship of creation and our Creator. For our Creator is not the lofty and distant designer so much as "the lamb that was slain before the creation of the world" (Rev. 13:8).

---

[7] W. H. Vanstone, *Love's Endeavour: Love's Expense* (London: Darton, Longman and Todd, 1977) 69-70.